SCHOLASTIC

The Great **BIG** Idea Book MATH

Dozens and Dozens of
Just-Right Activities for
Teaching the Topics and Skills
Kids Really Need to Master

New York • Toronto • London • Auckland • Sydney
Mexico City • New Delhi • Hong Kong • Buenos Aires

Teaching *Resources*

Special thanks to the teachers and other creative collaborators
who contributed ideas to this book (in alphabetical order):

Peg Arcadi, Cheryll Black, Stacey Brandes, Bonnie Bliss-Camara, Marianne Chang, Susan Coleridge, Eileen Delfini,
Gail Englert, Mitzi Fehl, Kelley Foster, Rita Galloway, Pamela Galus, Barbara Gauker, Trina Licavoli Gunzel,
Lynn Holman, Jim Kinkead, Dan Kriesberg, Sue Lorey, Becky Mandia, Carol Kirkham Martin, Marie Mastro,
Nanette Cooper McGuinness, Judy Meagher, Ruth Melendez, Janice Reutter, Kathy Rudlin, Charlotte Sassman,
Judi Shilling, Wendy Weiner, Judy Wetzel, Bobbie Williams, Wendy Wise-Borg,
Janet Worthington-Samo, and Elizabeth Wray.

Formerly published as part of the Best-Ever Activities for Grades 2–3 series:

Addition & Subtraction © 2002 by Deborah Rovin-Murphy and Frank Murphy

Multiplication © 2002 by Bob Krech

Measurement © 2002 by Bob Krech

Time & Money © 2002 by Deborah Rovin-Murphy and Frank Murphy

Graphing © 2002 by Jacqueline Clarke

Page 46: "Using Subtraction" by Lee Blair from *Arithmetic in Verse and Rhyme* (Garrard, 1971).

Page 127: "The Inchworm's Trip" by Sandra Liatsos from *Poems to Count On: 32 Terrific Poems and Activities to Help Teach Math Concepts*. Copyright © 1995 by Sandra Liatsos.
Reprinted by permission of Scholastic Inc.

Page 224: "Birthdays" by Sonja Dunn from *All Together Now*. Copyright © 1999 by Pembroke Publishers.
Reprinted by permission of Pembroke Publishers.

Throughout this book you'll find Web site suggestions to support various activities.
Please keep in mind that Internet locations and content can change over time.
Always check Web sites in advance to make certain the intended information is
still available and appropriate for your students.

Editor: Mela Ottaiano
Cover design by Maria Lilja
Interior design by Holly Grundon
Interior illustrations by Paige Billin-Frye
ISBN-13: 978-0-545-14701-9
ISBN-10: 0-545-14701-8

CONTENTS

CONTENTS

CONTENTS

Introduction

Welcome to *The Great Big Idea Book: Math*! Inside this book, you'll find dozens and dozens of activities, all of which will enrich your math program and support the National Council of Teachers of Mathematics (NCTM) standards.*

Many of these activities were contributed by teachers from across the country. The engaging activities provide opportunities to teach and assess math skills in fun and creative ways. They are drawn from all areas of the curriculum and many integrate more than one skill from across the disciplines. The activities also provide opportunities for students to work individually, in small groups, and as a class.

Highlights of the book include:

- ideas and activities from teachers across the country

- support for the many ways your students learn, including activities that link math with writing, art, music, poetry, and movement

- activities that support the NCTM standards

- hands-on science connections

- strategies for English language learners

- test-taking and assessment tips

- interactive displays that encourage children to collaborate in their learning

- computer connections, including software and Web site suggestions

- ready-to-use take-home activities and graphic organizers

- literature connections

- interactive and skill-building morning message ideas

- take-home activities to involve families in student learning

- and more activities to bring even greater effectiveness to your math program as you make math easy and fun!

* For more information see *Principles and Standards for School Mathematics*, published by NCTM (2000).

Addition & Subtraction

Children explore addition and subtraction concepts every day. They add and subtract coins to pay for lunch, they add and subtract as they share candy or cards, and they add and subtract as they keep score in games.

The activities in this chapter invite children to take their math explorations further, in ways that are just as appealing as playing a game or sharing a treat. Whether it's a lively game of Musical Math Chairs (see page 12), a math-filled morning message (see page 17), or history-making math stories (see page 37), children will gain confidence in their mathematical abilities and learn important skills as they participate in fun, hands-on activities.

All of the activities support the NCTM standards for numbers and operations, strengthening skills and concepts in the following areas: understanding numbers, ways of representing numbers, and relationships among numbers; understanding meanings of operations and how they relate to one another; computing fluently and making reasonable estimates; having number sense; estimating a reasonable result for a problem; making sense of numbers problems and results; solving problems using relationships among operations; and performing computations in different ways using different skills, including mental calculations, estimation, and paper-and-pencil calculations.

Though the activities are organized by addition and subtraction, most of them feature interdisciplinary connections. For example, Math Magic (see page 14) connects social studies and math as children learn about a puzzle that Ben Franklin invented more than 200 years ago. Math Art (see page 35) encourages art appreciation at the same time it involves children in practicing addition and subtraction skills. Hopscotch Subtraction (see page 26) integrates movement with math as children practice counting back. These interdisciplinary uses are engaging for kids and help demonstrate the importance of math in their daily life.

Picture the Problem

Tap into visual and spatial learning with an activity that combines fact practice with art.

Cover children's desks or tables with white roll paper. Tape it in place. Give children crayons or markers. Share an addition problem, and invite children to represent it by drawing pictures of their choice. They'll practice addition skills and have some fun drawing at the same time (often a very appealing and stress-free activity for children, which might well lead to better attention and retention of the math they're learning!).

TIP

When children are working with multiple-digit numbers, encourage them to write a regrouped number with a colored marker or write a small addition sign next to the digit to help them remember to add it!

Literature LINK

Anno's Counting Book

by Mitsumasa Anno (Crowell, 1975)

This classic, wordless picture book is perfect for introducing counting. While the pictures seem simple at first, with each turn of the page the counting becomes more complicated. Starting with zero and working up to twelve, there are many things to see on each page. Children will love finding and counting various things on each page that correspond to each number. An explanation at the back of the book offers a history of numbers.

Button Count

Before children can add multiple-digit numbers that require trading, they must understand the concept of trading 10 ones for 1 ten. Try this fun activity to reinforce this concept.

(§) Give each child a small paper cup of buttons. Have children take out one button for each button on their clothing that day.

(§) Assign partners to children. Have partners combine their buttons and then use manipulatives (such as "tens sticks" and "ones cubes") to represent their buttons in tens and ones.

(§) Bring students together to repeat the activity on a larger scale. Have children keep their tens bundles but pool their ones to see how many more tens they can make. Have them trade each group of 10 ones cubes for 1 tens stick. Count by tens together to see how many buttons the class is wearing!

To assess children's understanding of ones and tens, invite them to draw seven blocks, then draw eight more blocks. Have them ring a group of ten to show how to group tens and ones.

Twenty Wins

Invite students to practice adding four digits (and squeeze some subtraction in, too) with this challenging two-player game.

(§) Make copies of the reproducible grid on page 38. Give one to each child.

(§) Have children play in pairs, taking turns rolling two number cubes. Players can choose to add or subtract the two numbers on the cubes, then write the sum or difference in any box on the grid.

(§) A player scores a point when he or she can successfully put four digits together into a square (four boxes that form a square) to total 20. Students can use a highlighter or crayon to mark successful squares on the grid. The game continues until one player has received a given number of points or until the board is full of numbers.

Wendy Wise-Borg
Rider University
Lawrenceville, New Jersey

House of Addition

Invite your students to build a fun-filled house of addition and open lots of doors (and windows) to fun math-fact practice.

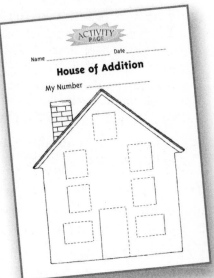

Give each child a copy of the house template on page 39. Have students fill in a number (at the top) and write different numbers on the window flaps and front door. The number at the top will be the missing addend for the numbers on the house (each combination, then, making a number sentence). Have students cut the windows and door into movable flaps and then glue the house to a sheet of construction paper, placing glue around the edges but not behind the movable flaps. Have students write the sum of each problem behind each flap. Display the houses of addition (flaps closed) and let students practice their math facts by solving the problems on each house and lifting the flaps to check their answers.

Bobbie Williams
Brookwood Elementary
Snellville, Georgia

For a challenge, have students select the number that will serve as the missing addend and record it on the back of the paper, not in the space provided. When you display the houses, children will have to determine the missing addend by looking at the numbers on the windows and door and the answer behind each flap.

Literature LINK

Domino Addition
by Lynette Long (Charlesbridge, 1996)

This colorful picture book uses dominoes to teach the concept of addition. Sets of dominoes on each colorful page represent sums from zero to twelve. Students can strengthen their skills by adding them up. After sharing the book, invite children to use real dominoes to create and practice addition number sentences.

Roll the Number Cubes

Invite pairs of children to play this quick and easy game together to create and complete two-digit addition problems.

You'll need two number cubes (preferably one cube with the numbers 1 to 6 and another with the numbers 4 to 9), a pencil, and paper. One child rolls the cubes and the partner writes a two-digit number from the cubes. So, for example, if a child rolls a 1 and a 3, the other child could write 31. The first child rolls again, creating a second number that the partner writes underneath. The partner completes the addition and the first child checks the work. Children switch places and play again.

Counting on Buildings

Invite children to build structures out of ones cubes and tens blocks (and hundreds blocks if available)—and build mental math skills, too.

Encourage children to count up the value of the cubes and blocks as they use them and to record the "sum" of their structure. Providing children with free time to explore building with these blocks will help them gain familiarity with the amounts. As a challenge, invite children to determine the sum of the cubes and blocks before they build. Have them record this number on a slip of paper. Classmates can estimate the sum represented in the structures, then see how close they came.

Literature LINK

The Grapes of Math

by Greg Tang (Scholastic Press, 2001)

This award-winning picture book uses brightly colored illustrations to introduce creative techniques for solving math problems. Described as "mind-stretching math riddles," the text is written in verse and will help children look at math in completely new ways.

Musical Math Chairs

Incorporate movement, music, and math with this skill-strengthening activity. You'll need a radio (or tape or CD player and music).

On each child's desk, place a pencil and a sheet of addition problems. Use various worksheets to provide experience with different kinds of problems. For example, at one child's desk, place a sheet of word problems. At another desk, place a sheet of one-digit addition problems (for example, 3 + 5 = __). You can also provide activity sheets for two-digit addition problems, problems that invite kids to find the missing addend, and so on. Tell students that they're going to help one another solve all of the problems on their sheets. Then start the music and have students get up and walk (or dance!) around the room while the music is playing. Stop the music and have students sit in the closest seat. Have each child solve a problem on the paper at that seat, then start the music again and repeat the activity. Continue until all papers have been completed! In the process, each child will have had practice trying different kinds of problems.

When learning about regrouping with addition, allow students to use a highlighter to mark the ones column before they solve a problem. This will visually remind them to begin with the ones column and to move the regrouped number they just added into the tens column.

Stick to It!

Every classroom can always use new math manipulatives. Invite students to make these artistic manipulatives while learning about tens and ones.

- Give children glue, craft sticks, and various trinkets, such as plastic beads, dried beans, and stickers.

- Have children arrange the objects on the sticks in tens and glue them in place. Children can sign their names to the backs of the sticks with phrases such as "Created by Diego" or "Built by Amy." Let the sticks dry, and use them for future math activities. For example, have students line up the craft sticks on a board tray one at a time and count by tens. How many objects all together? Can they represent this in a number sentence? Go further and use the sticks to introduce multiplication (10 x [the number of students] = [the total number of objects]).

Sum Stories

One of the NCTM standards is the ability to communicate mathematically. Here's an interactive display that can help students meet this standard.

- Display a white board and above it write the title "Sum Stories." Add a decorative border and write a number sentence at the top.

- Invite a child (or two children) to write an addition story about the number sentence. Children can draw pictures to go with the story, then leave it on display for the day to share with classmates.

- Each day, erase the board, change the number sentence, and let new children write and illustrate a story.

Judy Meagher
Student Teacher Supervisor
Bozeman, Montana

TIP

Using Braille flash cards is a great way to incorporate kinesthetic learning for all children. Check these Web sites for these materials and more: **www. cal-s.org** and **www.uncle goose.com**.

ENGLISH Language LEARNERS

To help English language learners better understand the addition and subtraction skills you're teaching, invite several children to act out a word problem that a child has written. The entire class will enjoy the acting experience, and English language learners will be able to understand the math in new ways.

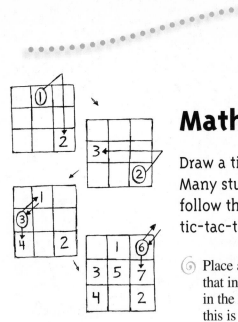

Math Magic

Math Magic

Draw a tic-tac-toe grid on the board. Ask students what it is. Many students will recognize the familiar game board. Then follow these directions to amaze your students with a little tic-tac-toe math magic! (See sample magic square, left.)

⚬ Place any number in the center box of the top row. Using numbers that increase or decrease consecutively, place each remaining number in the box that is one up and one to the right of the one before it. If this is not possible, use one of the following alternatives:

- If the next move is on top of the square, go to the bottom of that column and write the number in that box.

- If the next move is to the right of the square, go to the far left of that row and write the number in that box.

- If the next move is in a box that already has a number or is on top of and to the right of the whole square, go to the box directly below the previous one and write the number in that box.

⚬ Ask students if they notice anything about the numbers in the squares. List all responses, then draw another tic-tac-toe grid on the board.

⚬ Now, share *Ben Franklin and the Magic Squares*. (See below.) When you get to the section about Ben Franklin creating the magic square, use the blank tic-tac-toe grid on the board to model the placement of the numbers. Review the addition inside the square, across, down, and diagonally. Then give children copies of the reproducible tic-tac-toe grid on page 40 and invite them to create their own magic squares.

Literature LINK

Ben Franklin and the Magic Squares
by Frank Murphy (Random House, 2001)

This easy reader combines history and math and serves as a great springboard for creating really interesting math puzzles. It is a true story about Ben Franklin and how he made the math puzzles called magic squares.

TIP

Challenge students with a "problem of the day" that consists of a magic square with a few missing numbers. Students have to find the magic sum, then find missing addends to complete the square—great practice for using multiple strategies and adding with three digits.

Magic Triangles

Ben Franklin's squares aren't the only shapes that can be magical. Invite your students to pair up and have fun strengthening addition skills with this challenging problem.

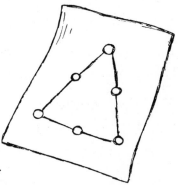

Draw a triangle on the board. Draw a circle at each angle and in the middle of each line. (See sample.) Note that the lines do not pass through the circles. Have students copy the triangle on paper or mini-chalkboards. Then challenge them to arrange the numbers 1, 2, 3, 4, 5, and 6 in the circles so that the sum of each side of the triangle is the same. After children complete the triangle, compare answers to see if they are correct and to see if there is more than one possible arrangement. Ask children to think about what would happen if each digit were increased by 10, 100, or 1,000. Make new magic triangles and try it out!

Peg Arcadi
Homeschool Teacher
Trumansburg, New York

The Greatest-Sum Game

Invite students to improve their number sense and addition skills with a game that challenges players to make an addition problem with the greatest sum.

- ⑤ Have children pair up to play. Give each player a copy of the game board on page 41.

- ⑤ Each pair of children will need a set of the digit cards on page 42. Have students cut out the digit cards, shuffle them, and place them facedown between players.

- ⑤ Players take turns taking a card and writing that number in any box. Players continue taking turns until all boxes have been filled.

- ⑤ Now each player adds the numbers in each set of squares to complete the addition problem. Have players circle their greatest sum and then exchange papers to check each other's addition. Players then compare their answers. The player with the greatest sum wins.

Change the game to make an addition problem with the smallest sum. Further challenge students by making the object of the game to have a sum come closest to a certain number without going over.

Math Connections Word Wall

Help your students make math connections in the world around them by building an illustrated word wall.

Invite children to think of ways to represent ten or one hundred. Encourage them to be creative—for example, *decade*, *dime*, a ten-dollar bill, ten-yard line, ten pennies, and the Roman numeral X (for example, on an old building or in the page numbers of a book) all represent ten. *Century*, a hundred-dollar bill, the Sacagawea coin, the Roman numeral C, and a football field all represent one hundred. Write ideas on sentence strips and display to make a word wall. To reinforce the concept of ten and one hundred in the real world, have children draw and display pictures to illustrate the words.

Literature LINK

Counting on Frank

by Rod Clement (Gareth Stevens, 1991)

In this wild story, a young boy and his dog, Frank, examine everything from peas to ballpoint pens, experiencing all kinds of counting fun. The boy's imagination allows readers to learn challenging ways to calculate some weird things. A guide in the back gives hints and answers to problems.

TIP

Remind students about the importance of aligning multiple-digit numbers when adding and regrouping. Give students problems written horizontally on chart paper or the board. Invite students to write down the problems on grid paper or lined paper turned sideways. Using these papers will help students position the numbers in the correct place-value column.

Morning Message Math

Use your morning message as a way to reinforce addition and subtraction skills every day. Here are a few ideas to try:

- Ask students to see if the day's date is the sum of a double—for example, if the date is the 24th, this is the sum of a double (12 + 12).

- Invite students to write an addition or subtraction sentence that has the day's date as the answer.

- Let students count the number of words in the morning message by twos.

- Incorporate a sentence in your morning message that asks students to add or subtract to find an answer—for example, "Monday it was 73°. Tuesday it was 84°. Which day was warmer? By how much? How hot do you think it will be today?" Students can record answers beneath the morning message.

Visit an outstanding math Web site (for kids and teachers) packed with games, activities, and more at **www.cool math4kids.com**.

Nine Digits Equal One Hundred

Challenge students to add numbers to make 100 with this mind-bending activity.

Write the sequence of digits 1–9 on the board. Ask children to copy the digits, leaving a space between each digit. Invite children to place plus or minus signs between some spaces so that the answer to the number sentence will be 100. (Remind them that the numerals have to stay in the same sequence.) Some possible solutions are:

$$1 + 2 + 3 - 4 + 5 + 6 + 78 + 9 = 100$$
$$12 - 3 - 4 + 5 - 6 + 7 + 89 = 100$$
$$123 + 4 - 5 + 67 - 89 = 100$$

$$12 + 3 - 4 + 5 + 67 + 8 + 9 = 100$$

Bob Krech
Dutch Neck Elementary
Princeton, New Jersey

Count On to Win

Invite children to practice the skill of "counting on" with number and dot cubes.

Provide a number cube (1–6) and a dot cube (1–6). Let pairs of students take turns rolling both cubes. When a player rolls a 1, 2, or 3 on the dot cube, the player says the number on the number cube, then counts on the number of dots to find the sum. Players can keep a tally of how many times they got to "count on" (how many times the dot cube landed on 1, 2, or 3). The player who gets to count on the most times wins.

Literature LINK

Mission Addition

by Loreen Leedy (Holiday House, 1997)

This picture book is presented in the form of comic strips. Hippo teacher Miss Prime and her students practice all kinds of addition concepts while solving various problems that have them adding columns of numbers, creating addition sentences, and adding dollar amounts.

ENGLISH Language LEARNERS

The concept of counting on will be unfamiliar to most English language learners. Explain that to count on means to start with a certain number and count by ones to add. (Usually children are taught to use this strategy when adding 0, 1, 2, or 3.) Ask children to think of other ways they have heard the word *on* used—for example: "The sock is on my foot" or "Turn on the lights." List these examples on the board. Invite students to draw and color pictures showing different meanings of the word *on*, including the phrase *count on* in math.

___ Ways to Get to 11 Class Book

To reinforce different ways to make a sum, make a class book modeled on the story *12 Ways to Get to 11*. (See below.)

- Take time to review the story. Which was your students' favorite way of adding up to 11? Which was the most unusual? The funniest?

- Ask children to suggest other ways to get to 11. Then give each child a sheet of drawing paper. Have children make up and illustrate a new way to get to 11.

- Put the pages together to create a class book. Add a cover with the title "___ Ways to Get to 11."

- Fill in the blank with the number of kids in your classroom, then bind the pages and let children take turns reading aloud their pages and sharing their illustrations.

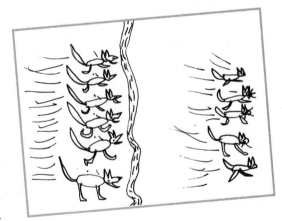

Literature LINK

12 Ways to Get to 11

by Eve Merriam (Simon & Schuster, 1993)

"One, two, three, four, five, six, seven, eight, nine, ten . . . twelve. Where's eleven?" This question starts the quest for different ways to add up to eleven. From counting six peanuts and five pieces of popcorn at the circus to counting a sow and ten baby piglets, children have the opportunity to investigate colorful illustrations while learning about different ways to make a sum.

TIP

For a Web site full of great addition and subtraction games, go to Mrs. Dowling's Fun and Games Page: **www.dowling central.com/ MrsD/fun.html**.

Reflecting Doubles

Invite your students to visually enhance their understanding of doubles by using a mirror. The effect is magical, and
students will want to do it again and again.

⑥ Give students different quantities of small items like candies,
buttons, or jacks. Have students place them on their desks. Ask
students if they can think of a way to double the number of items
on their desks without getting any more.

⑥ Hand out small mirrors. Ask students to predict what will happen
when they hold the mirror right up against the items. Let students
try it out and then count the items they see.

⑥ Have them write number sentences to show the "doubles" math,
then try magically doubling other items on their desks.

Literature LINK

Two of Everything

by Lily Toy Hong (Whitman, 1993)

In this Chinese folktale, an old couple finds a magic pot that doubles everything that is put into it. The trouble starts when the old
wife falls into the pot! After sharing the book, reinforce the concept of adding doubles by bringing in a large plastic cauldron.
Fill the cauldron with everyday items. Prepare to amaze the kids
by secretly placing a set number of items in the cauldron and
then handing out the matching number of items to students. Ask
students to place their items in the cauldron. Then magically pull
out the item—now doubled!

Mail Math

Students practice real-life math skills while combining different numbers to create a set amount of money.

⑤ Make copies of the reproducible stamps on page 43. Distribute blank envelopes along with a set of stamps to each student.

⑤ Have students write their own names and addresses on the envelope, and explain that letters weighing more than a certain amount require different amounts of postage.

⑤ Select one student to be Postmaster General. Have that student stand at the front of the room and announce the amount of postage needed to mail their letters.

⑤ After calling out the amount, have students combine stamps on the envelope to make exactly the amount they need. Let students share the different combinations they used. Record them on the board, along with the total. Then select another student to be the Postmaster General and give a different total to try.

TIP

Use everyday classroom procedures to observe students' proficiency with addition and subtraction. Lunch count, attendance, paper passing, and other routines all lend themselves to real word-problem solving.

ENGLISH Language LEARNERS

Picture clues are a great way to have your English language learners make connections between the concrete action and abstract vocabulary and symbols used to solve problems. Post pictures of a set with an action. (For example, display five candy bar wrappers and put an X through three of them.) Underneath the pictures, write the subtraction sentence in both symbols and words—for example, 5 - 3 = _____ and "I had five candy bars and ate three. How many do I have left?"

Take a Turn

Children love playing games. Capitalize on this interest by using familiar games to practice math facts.

First, choose games that require players to pick a card in order to go forward. (Chutes and Ladders and Candy Land are good examples.) Substitute math fact flash cards (your students can make them on index cards) for the original game cards. Students need to give the correct answer to the math fact problem to find out how many spaces to move forward. This is an easy way for teachers to create games for centers as well as a useful way to recycle old game boards.

Wendy Wise-Borg
Rider University
Lawrenceville, New Jersey

Calendar Subtraction

Have some fun figuring out how many school days there are each month and strengthen subtraction skills at the same time.

- Begin by inviting students to suggest ways they could find out how many school days there are this month—for example, they could count up all the weekdays and subtract the number of school holidays that fall during the week or subtract the number of weekend days and holidays from the total number of days in the month.

- Try out one of the methods together and write the resulting number sentence and answer on a sentence strip.

- Have students work together to write a few sentences to go with the number sentence, explaining what it shows. Display both next to the calendar. As you repeat this activity each month, have children make comparisons: "Which month has had the most school days so far? The fewest?"

Towering Comparisons

Children love to compare sizes of some of the world's tallest structures. Strengthen their ability to find differences by supplying them with data to compare.

Display information about the Empire State Building, Sears Tower, and Petronas Tower. (See How High?, right, for data.) Have children use the data to calculate the differences between the height of the various structures. Use the activity to highlight the importance of recognizing that subtraction does not always mean taking away—it can mean comparing. To go further, let students research other tall structures to compare. Invite them to illustrate their results and display them on a bulletin board or in the pages of a class book.

Peg Arcadi
Homeschool Teacher
Trumansburg, New York

How High?
Petronas Tower:
1,476 feet

Sears Tower:
1,454 feet

Empire State Building:
1,250 feet

Race to Zero

Students will have fun playing this backward game while reinforcing and practicing their subtraction skills.

Divide the class into groups of two or three students. Each group will need a copy of the game board on page 44, a number cube, and a marker for each player. Have players put their markers on the space marked 30. Players take turns rolling the number cube and subtracting that number from their position on the game board. The first player to get to zero wins the game!

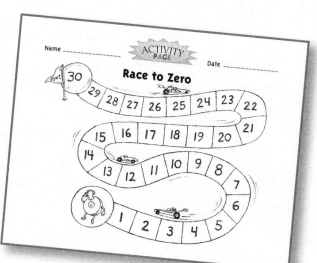

Make paper, pencils, and calculators available to check each player's accuracy.

Math Snack

Students will enjoy math even more if they know they can eat the problems when they are done! Plan a math snack picnic to reinforce combining different-sized sets.

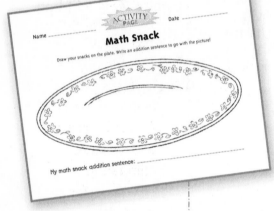

- Invite families to send in snack foods to share. Items that come in small pieces work well—for example, pretzels, carrot sticks, crackers, and grapes.

- Arrange the snacks on a table, along with plates, eating utensils (if necessary), and napkins. Have students select foods they'd like for snack and place them on the plate.

- Before they take their first bite, give each child a copy of page 45. Have children draw pictures of their snacks and then write addition sentences to go with them—for example, "5 carrots + 12 grapes + 3 crackers = 20 snacks."

- Strengthen subtraction skills by telling students to eat their snacks! Have them pause periodically to share math sentences with one another about what they're eating and what's left—for example, "five carrots take away one carrot equals four carrot sticks."

Peg Arcadi
Homeschool Teacher
Trumansburg, New York

TIP

Check for food allergies before serving snacks.

Literature LINK

The Candy Counting Book: Delicious Ways to Add and Subtract
by Lisa McCourt (BridgeWater, 1999)

This colorful picture book uses all kinds of classic candies and delicious story problems to help kids learn about addition and subtraction. Let your students think of their own delicious ways to add and subtract. Use them to make a mouth-watering display!

Subtraction Action Word Wall

Strengthen students' understanding of subtraction and the language involved by sharing real-life subtraction situations and using them to build a word wall.

- Read the following situation to children (or make up your own using students' names): "John had 24 sheets of paper. He gave 12 sheets to Nico. How many sheets of paper did John have left?"

- Ask children to discuss the actions involved in these situations. (*giving away* and *comparing*) Ask: "Which words show the subtraction action?" (*gave* and *have left*)

- Repeat the activity with new situations. Incorporate subtraction words such as *have left* and *how many more than* to show the different operations in the subtraction problems.

- Follow up by asking students to volunteer additional "action" words for subtraction problems—for example, *how many more, how many less, greater than, less than, have left, have now,* and *have before.* List students' responses on sentence strips or poster-board to make a Subtraction Action Word Wall.

Subtraction Straws

Students break apart bundles of straws to make the function of regrouping more visual and hands-on.

Divide the class into small groups. Give each group a handful of straws and some trash-bag ties. Have children use the ties to bundle straws in groups of ten. Ask them to leave at least nine straws unbundled. When students are ready, give each group a subtraction problem to solve. Have students match the number of straws with the problem and decide whether or not they'll need to untie and break apart a tens bundle. So, to show 12 − 9, students would start with a bundle of ten straws plus two single straws, then break apart the tens bundle to subtract nine.

TIP

Many children say "take away" for subtraction signs—for example, "four take away two." Remind them that subtraction also includes comparing numbers—for example, comparing children's height at the beginning and end of the year to see how much they've grown.

Hopscotch Subtraction

Take subtraction out to the playground by playing this variation of hopscotch.

Draw a hopscotch lane, or use the ones already painted on your playground. Make a giant number cube by covering a cube-shaped tissue box with construction paper and writing the numbers 1 to 6 on it. Let students take turns rolling the number cube and, starting at 10, hopping back (subtracting) the number that appears on the cube. The goal is to hop back (in as many turns as it takes) until they reach 1 exactly. Players skip turns if they roll a number that would result in going beyond the 1 on the hopscotch lane.

Subtraction Poetry

Get your students excited about subtraction with writing and poetry!

Give each child a copy of page 46. Read aloud the poem "Using Subtraction." Invite students to fill in the blanks with things they would like to see subtracted from their world—for example, they might fill in such pesky problems as homework or rainy days, or they might suggest more serious concerns, such as poverty or pollution. Invite students to brainstorm more ideas, and have them fill the blanks with as many things they wish they could subtract.

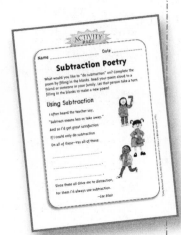

Literature LINK

Subtraction Action
by Loreen Leedy (Holiday House, 2000)

A hippo teacher named Miss Prime and her class learn about subtraction at the school fair. From cookies disappearing in a puppet show to prices being reduced at refreshment stands, this picture book is a great springboard for introducing subtraction concepts. Colorful illustrations provide visual clues to help with the problems.

The Greatest Difference Game

This game reinforces number sense and gives children practice subtracting three-digit numbers. You can adapt it to use with more or fewer digits.

- You'll need three number cubes and each child will need paper and pencil. (The game will work best with cubes that have the digits 0–5, 2–7, and 4–9.) Students can play in pairs or more.

- A player rolls three number cubes. Both (or all) players use the three digits to write a three-digit number. Each player should hide the number he or she writes.

- Another player rolls the cubes again. Both (or all) players make another three-digit number and write it down. Each player then subtracts the lesser number from the greater number. Each player's difference is the score he or she uses for that round.

- Players continue taking turns and recording each three-digit number and score for the round on a sheet of paper. The first player to reach a specified number (1,000 is a good start) is the winner. After playing, invite children to discuss any strategies they used. For example, ask students why and when they may have written the highest or lowest number possible.

Jim Kinkead
Newtown Elementary
Newtown, Pennsylvania

Make extra copies of the Make a Math Alien activity sheet and complete them with problems you want children to solve. Place the papers at a math center and let children create more creatures while strengthening their addition and subtraction skills.

Make a Math Alien

Your students will use their imagination while practicing out-of-this-world addition and subtraction facts.

Give each child a copy of page 47. Have students fill in the blanks without solving the equations—for example, "5 + 2 = _____ eyes." Let children exchange papers to solve each other's equations, then use the answers (*7 eyes, 11 heads,* and so on) as directions to create math aliens. When children have finished creating the aliens, have them share their out-of-this-world art with the classmate they traded papers with.

Problem-Solving Stories

Integrate language arts with mathematical problem-solving by having students write literature-inspired word problems.

Invite students to think of a character and event from a favorite book. Model for students how to turn both into a word problem—for example, "Curious George has to gather 12 bananas for the man in the yellow hat. He already has 3. How many more does he need?" Have students write and illustrate a word problem that features their character and event. Students can place their stories in the matching books for other students to solve.

School-Days Subtraction

Get your students excited about summer, whether it's late or early in the year!

Tell students how many school days there are in the school year and display this information near the classroom calendar. Ask students how many days they've been in school. They can use a school calendar to find out. Now invite them to use their three-digit subtraction skills to find out how many school days are left in the school year. Keep a running count each day or each week and post it alongside the calendar.

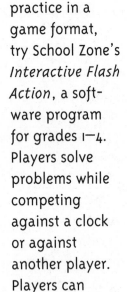

Literature LINK

Shark Swimathon: Subtracting Two-Digit Numbers

by Stuart J. Murphy (HarperCollins, 2001)

A shark swim team practices two-digit subtraction while trying to reach a goal of 75 laps. The subtraction gets more difficult as the story progresses, and a swim team coach is on hand to explain the process in each example. Perceptive children will note that the swimmers' performance improves with practice!

Choral Countdown

Many students use the skill of counting back and make the mistake of starting with the number that will be decreased. Strengthen your students' skills in counting back by regularly doing this choral countdown.

TIP

Challenge children by setting the number to be subtracted or the number to be counted back (such as 9 or 10). Say the starting number. Have students then call out the number that is one less and continue counting back. What number will students count back to?

◉ Before you begin a choral countdown, give students some practice in starting with the correct number. Say a number, and have students call out the number that is one less than that number. Repeat this quite a few times to make sure students understand the concept. Then try a choral countdown.

◉ Announce a number and tell students what number you want them to count back to. On your cue, have students start counting back together. To keep track of how many numbers they're subtracting, have them make tally marks on dry erase boards or mini-chalkboards.

◉ After arriving at the target number, they can add up the tally marks to determine the number they subtracted from the original number.

ENGLISH Language LEARNERS

U sing a large calendar to practice counting back makes the activity more visual for English language learners. Point to the numbers on the calendar as students count them back aloud. Seeing the numbers will give students another way to remember the words and will build not only number sense but vocabulary, too.

Round and Round Comparisons

In this activity, students move to music and get lots of practice with subtraction skills in the process.

- Divide the class into two equal groups. Distribute multilink cubes or other manipulatives to each student. Vary the number of manipulatives you give each child (for example, from 1 to 20). Have one of the groups form a circle. Have the other group form a larger circle around the first group.

- Play music and have children move around in their circles. Both circles can go in the same direction, or you can have the circles move in opposite directions. When you stop the music, have each student in the outer circle partner up with the student across from him or her in the inner circle.

- Have outer/inner circle partners match up their multilink cubes and use subtraction to compare who has more and who has less. After students have made their comparisons, put on the music and play again.

It's in the Bag

Students use critical thinking and practice finding missing addends while figuring out the number of hidden objects in a bag.

- Distribute manipulatives and a paper bag to pairs of students. The manipulatives can match a theme students are studying—for example, if the theme is explorers, you might use gold coins. If you're exploring oceans, you might use fish-shaped crackers. To connect with a unit on bugs, use plastic insects.

- Have one student (A) in each pair set out a certain number of manipulatives for the other child (B) to see and count. The remaining manipulatives should be placed to the side.

- Have partner B close his or her eyes while partner A puts some of the manipulatives in the bag. Partner B then has to guess how many objects are in the bag by counting the number of objects that are out of the bag and using addition or subtraction to figure out the missing addend. Have partners switch places and play again.

TIP

Strengthen students' skills with *Mega Math Blaster* (Davidson), a software program packed with arcade-style games that reinforce relationships among numbers and meanings of operations and build fluency in computing.

Who Am I? Math Riddles

Do your students love to create riddles? Invite children to choose a favorite number to create lift-the-flap subtraction math riddles for a friend.

Make copies of the reproducible Who Am I? riddle sheet on page 48. Have students cut out a square of construction paper to fit over the answer space. They can make a flap to hide their answers by gluing the top edge of the construction paper square to the top edge of the answer space. Have children complete the riddle for any number they like and then lift the flap and write the answer in the space provided. Display the riddles on a bulletin board or in the hallway and invite classmates or other children in the school to solve the riddles and lift the flaps to check their answers.

First off the Bridge

Students can practice counting on and counting back as they try to be the first person off the bridge in this game.

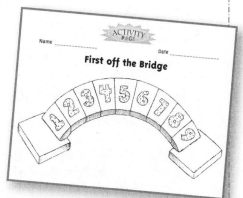

Have students pick partners. Give each pair of students a game board (see page 49), eight index cards labeled "ADD 1," "ADD 2," "ADD 3," "ADD 0," "SUBTRACT 1," "SUBTRACT 2," "SUBTRACT 3," and "SUBTRACT 0," and two game markers (such as dried beans). Have students place their markers in the middle of the bridge and the cards facedown in a pile. To play, children take turns choosing a card from the pile and either adding (going forward) or subtracting (going backward) from their position on the bridge. The goal is to be the first person who gets off the bridge. Players do not have to land exactly on the first stone off the bridge to win. The player can end up multiple steps off to win.

Becky Mandia
Newtown Elementary
Newtown, Pennsylvania

TIP

Place additional copies of the First off the Bridge game, along with simple directions, in resealable plastic bags. Let children take turns borrowing the games to play with their families. Include a response sheet in the math packs so that families can share comments and strategies.

All Through the Year

At the start of a new month, invite children to combine math and science to learn more about the weather.

- Set up a schedule to have children observe the weather each day.

- Make a chart to record whether the day was sunny, cloudy, rainy, or snowy.

- Have children use tally marks to record the information on the chart. They can also use words or pictures to record the same information each day on the calendar. At the end of the month, have children work in groups to discover differences between the numbers of sunny days and of rainy days, cloudy days and sunny days, and so on. Remind children to add the tally marks first, then subtract the numbers to find differences. Discuss the findings, and post them next to the calendar.

Mix up the puppets and place them at a center. Let students work with partners to reorganize the families and put on mini puppet shows that incorporate the math.

Fact-Family Puppets

Help children understand relationships between addition and subtraction by creating a puppet "family of facts."

- Invite students to write a simple math equation on a sheet of paper. Explain that each of their math equations is part of a fact family. Each member of this family has the same three numbers—for example, $2 + 3 = 5$, $3 + 2 = 5$, $5 - 2 = 3$, $5 - 3 = 2$. Have students write down the rest of the members of their fact families on the sheet of paper.

- Give each child four large craft sticks. Have students use a marker to write each of the fact family member equations at the bottom of a craft stick.

- Now for the fun part. Give children various craft supplies, such as yarn, glitter, felt, stickers, and colored stones. Let children use the materials to create each "family member." When they're finished, let students introduce their family to the class.

Fact-Family Jumping

This lively game reinforces relationships among numbers and helps children learn their math facts!

- Write the digits from an addition or subtraction sentence on paper plates. For example, write the numbers 3, 4, and 7 on three plates (one number per plate). Place the plates on the floor (numbers facing up) to form a triangle.

- Model fact-family jumping with addition by asking a student to stand on the 3. Have the student shout "Three" and then jump from the 3 to the 4. The student then shouts "plus four" and then jumps to the 7. The student shouts "equals seven" and then hops off.

- Try the same thing in reverse to model subtraction. Have the child start on 7 and then jump to either 3 or 4, then to the remaining plate for the answer.

- Let children make their own fact-family jumping games. (Check to make sure they're each doing a different one.) Let students rotate to one another's games to "jump" the fact families.

- Make this activity more challenging by turning over a plate and hiding a number. The student will have to shout out the missing number.

Add or Subtract? Morning Message

Strengthen students' critical-thinking skills with a morning message that combines addition and subtraction.

Display a certain number of manipulatives (magnets, stickers, and markers) next to your morning message. Clip a number card to the message (for example, the number 14 written on an index card). In your message, invite students to count how many manipulatives there are (for example, 10) and write that number in their notebooks. Ask them to decide whether they need to add or subtract manipulatives to get to the number on the card (for example, $10 + \text{or} - \_\_ = 14$). Have students write the problems in their notebooks. Let them share their answers and strategies at your morning meeting or math time. Let children demonstrate with the manipulatives what they did to solve the problem.

Rita Galloway
Bonham Elementary
Harlingen, Texas

Surprise children by using thematic manipulatives in the morning message math activity. You can order inexpensive seasonal items through Oriental Trading Company at **www.oriental trading.com**.

Hop to It

Students actively solve addition and subtraction problems as they hop up and down a giant number line.

○ Make a number line by taping a ten-foot piece of masking tape on the floor. Tape number cards (1–20) at six-inch intervals along the number line.

○ Write addition or subtraction problems on sentence strips, and you're ready to play! To play with the class, line up students and hold the number sentences facedown. Flash a sentence strip to the first student. Have this child hop out the problem on the number line to land on the answer. Remind students to always stand on the higher number first before jumping forward—for example, if the problem is 5 + 3, the student stands on the number 5 and hops three spaces forward to find out that the answer is 8.

○ Encourage students to count on or count back out loud as they hop up or down the number line to the answer. A fun variation of this activity is to give the students two numbers (for example, 3 and 9) and ask them how many hops it takes to get from one to the other.

Cheryll Black
Newtown Elementary
Newtown, Pennsylvania

Add/Subtract Graphic Organizer

This interactive bulletin board lets students access the correct terms when discussing math.

Ask students to name words or phrases that relate to addition and subtraction—for example, *equals, addend, difference*. List each word or phrase on a separate index card. Create a large Venn diagram on a bulletin board. Label one side "Addition" and the other "Subtraction." Let children take turns selecting a card and leading a discussion about where the card belongs. Have them place the cards in the correct section, either in the addition or subtraction section or in the overlapping section to show that it goes with both. As you continue to study addition and subtraction, add new cards. Students can use the board as a vocabulary reference when they're writing about or discussing the two operations.

Jim Kinkead
Newtown Elementary
Newtown, Pennsylvania

Laminate number line strips on each student's desk. Students can use these number lines as "helpers" to count on, count back, and skip-count.

Math Art

Math is everywhere! Invite students to explore math in their world by creating math problems to match a picture or drawing.

⑤ Gather assorted art prints, books with lots of art, pictures from calendars, and so on. Display them in a central location for students to browse.

⑤ Ask each child to choose a piece of art to explore. After giving children time to look closely at their work of art, challenge them to create word problems for mathematical equations that match the art. For example, a math sentence to go with Vincent van Gogh's *In the Bedroom* is: "How many chairs? How many chair legs altogether?"

⑤ Use students' word problems and the artwork to create a display. Take time to solve the problems as a class.

⑤ Encourage children to revisit the display on their own for more practice. Children will enjoy adding to the display with new math art.

How many legs altogether?

Literature LINK

A Child's Book of Art

by Lucy Micklethwait (Dorling Kindersley, 1993)

This oversized book features paintings and prints that encourage art appreciation while introducing math concepts, too, including numbers and shapes. It's a perfect resource for Math Art (see above), and includes works by Cassat, van Gogh, Matisse, Degas, Hiroshige, Renoir, Rivera, and others.

Around the World With Numbers

Enrich children's understanding of numbers and world languages by letting them create word problems using words or symbols for numbers in another language.

After sharing *Count Your Way Through China* (see below), and other books in the series, let children work together to make a word wall or poster listing the number names for 1–10 in each country's native language. Let children choose a favorite language and write a number sentence using the number words in that language. Have students arrange their word problems on a display. Add maps and flags to complete the display.

Literature LINK

Count Your Way Through China

by Jim Haskins (Carolrhoda, 1988)

Share this informative story to explore China and learn to count to ten in another language. Other books in the series introduce counting to ten in French, Spanish, Italian, and other languages. Use the information to create a picture word wall or display that shows how to count to ten in each language. For a related activity, see Around the World With Numbers, above.

Take-Home Activity:
Math Around Me

Students will enjoy working with families to come up with word problems connected to various places in their homes.

Give each child a copy of page 50 to take home. Go over the page together, reviewing the directions and letting students suggest possible problems to model the assignment. For example, students might suggest that they could label one room in the house "Kitchen," then write a word problem about cooking—for example, "If a cookie recipe calls for three eggs and there are 12 eggs in the carton, how many eggs will be left after I make the cookies?"

Word Problems From the Past

Connect history and math with word problems that take students back in time.

Model this activity by showing students an interesting historical fact that includes numbers—for example, "Clara Barton was 11 years old when she left school to care for her injured brother. She left a job at the U.S. Patent Office 30 years later in 1862 to help injured soldiers in the U.S. Civil War. How old was Clara when she became a Civil War nurse?" Invite students to write word problems using historical facts. Display the word problems on a bulletin board for classmates to solve.

Literature LINK

Fun With Numbers
by Massin (Harcourt Brace, 1995)

This picture book is full of interesting history about numbers. From why we count in sets of 60 with clocks to who used zeroes and who didn't, this book will fascinate young readers whether they are just learning to count or can add and subtract big numbers.

Name _____ Date _____

Twenty Wins

Name _____   Date _____

House of Addition

My Number _____

Name _____   Date _____

Math Magic

Name _____   Date _____

The Greatest Sum Game

Name _____   Date _____

The Greatest Sum Game

0	1	2	3	4	5	6	7	8	9
0	1	2	3	4	5	6	7	8	9
0	1	2	3	4	5	6	7	8	9
0	1	2	3	4	5	6	7	8	9
0	1	2	3	4	5	6	7	8	9
0	1	2	3	4	5	6	7	8	9
0	1	2	3	4	5	6	7	8	9

Name _____ Date _____

Mail Math

1¢	1¢	1¢	1¢	1¢
1¢	1¢	1¢	1¢	1¢
5¢	5¢	5¢	5¢	5¢
10¢	10¢	10¢	10¢	10¢
25¢	25¢	25¢	25¢	25¢
50¢	50¢	50¢	50¢	50¢

Race to Zero

30 start

29 28 27 26 25 24 23 22

21 20 19 18 17 16 15

14 13 12 11 10 9 8 7 6

5 4 3 2 1

The Great Big Idea Book: Math © 2009, Scholastic Teaching Resources

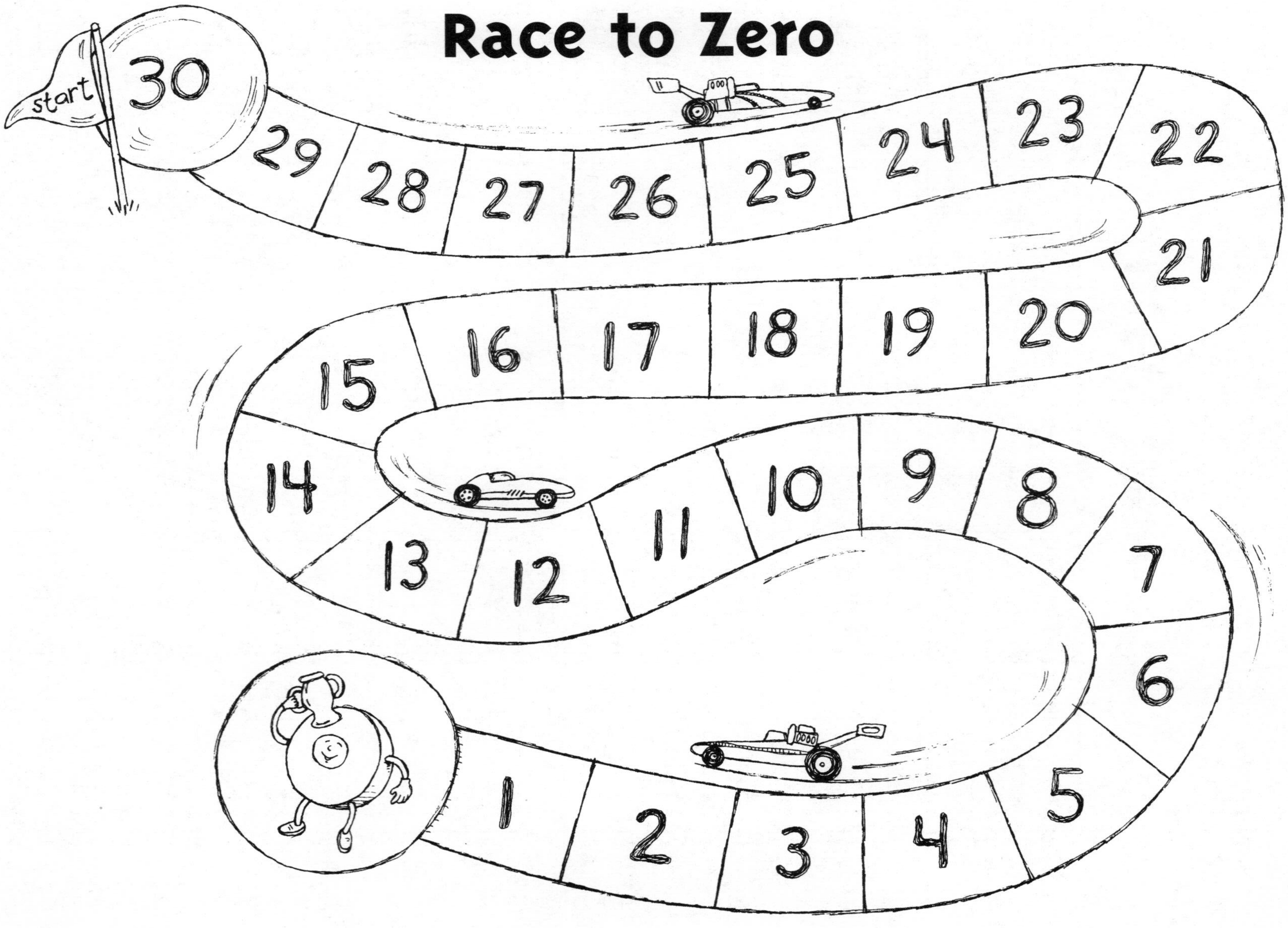

ACTIVITY
PAGE

Math Snack

Draw your snacks on the plate. Write an addition sentence to go with the picture!

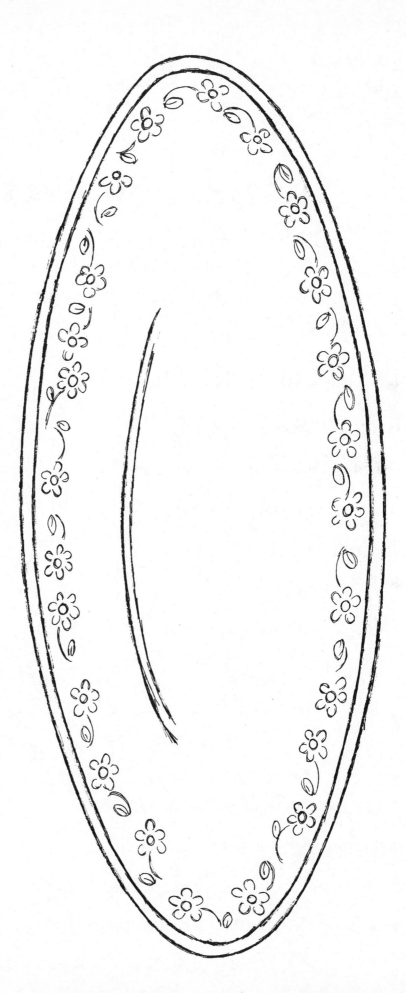

My math snack addition sentence: _____

The Great Big Idea Book: Math © 2009, Scholastic Teaching Resources

45

Name _____ Date _____

Subtraction Poetry

What would you like to "do subtraction" on? Complete the poem by filling in the blanks. Read your poem aloud to a friend or someone in your family. Let that person take a turn filling in the blanks to make a new poem!

Using Subtraction

I often heard the teacher say,

"Subtract means less or take away."

And so I'd get great satisfaction

If I could only do subtraction

On all of these—Yes all of these:

_____ ,

_____ ,

_____ ,

Since these all drive me to distraction,

For them I'd always use subtraction.

—*Lee Blair*

"Using Subtraction" by Lee Blair from ARITHMETIC IN VERSE AND RHYME (Garrard, 1971).

The Great Big Idea Book: Math © 2009, Scholastic Teaching Resources

ACTIVITY PAGE

Name ______________________ Date ______________

Make a Math Alien

Choose and write addends in the blanks. Trade papers with a partner. Complete each other's number sentences and draw the alien!

______ + ______ = ______ eyes

______ + ______ = ______ heads

______ + ______ = ______ legs

______ + ______ = ______ noses

______ + ______ = ______ ears

______ + ______ = ______ antennae

______ + ______ = ______ arms

______ + ______ = ______ hands

______ + ______ = ______ fingers

______ + ______ = ______ mouths

______ + ______ = ______ bodies

Name _____ Date _____

Who Am I?

When you double me you get _____.

When you double me and add one you get _____.

I'm one less than _____.

If you add ten to me you get _____.

Who am I? _____.

The Great Big Idea Book: Math © 2009, Scholastic Teaching Resources

Name _____

Date _____

First off the Bridge

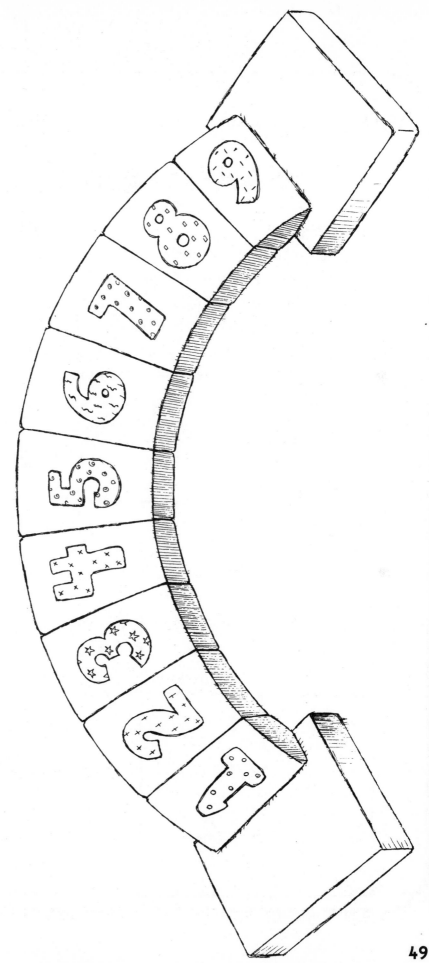

Name _____ Date _____

Math Around Me

Dear Family,

Your child has been practicing computation skills in class. Try this activity to strengthen these skills at home. Together with your child, choose four locations in your home and write them at the top of each square area in the house below. For each one, select some items that can be counted to make an addition or subtraction word problem. Write the problem in the appropriate space. Then solve it!

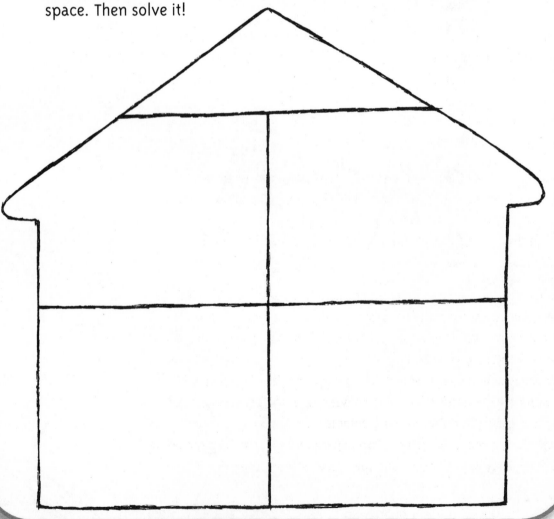

The Great Big Idea Book: Math © 2009, Scholastic Teaching Resources

Multiplication

Multiplication! Pippi Longstocking, the heroine of Astrid Lindgren's classic children's books, called it plutification and found it to be perplexing and a bother. Like Pippi, many of our own students are equally intimidated by the whole idea. Learning multiplication is almost a rite of passage in elementary school. It awaits

children somewhere around second or third grade. Rumors spread even among the kindergartners on the very first day of school that if you don't know your multiplication tables, you may as well go home now. They believe, from experience or otherwise, that multiplication consists solely of the memorization of an insurmountable number of facts, all waiting for them on an endless series of timed tests.

Of course, it doesn't have to be that way. And even if you can't help Pippi, you can provide your students with an exciting and interesting learning experience with multiplication. By providing motivating, meaningful experiences that help students gain real understanding of multiplication, this area of math can be made fun and easy.

Are learning the facts still important, though? Absolutely! The NCTM *Principles and Standards* (2000) states, "Knowing basic number combinations—the single-digit addition and multiplication pairs and their counterparts for subtraction and division—is essential." This is a reasonable goal that can be reached in ways that are engaging and developmentally appropriate. The activities are broadly arranged from introductory explorations to more complex applications. You can follow the sequence, or you can dip in here and there and adapt the activities to meet your needs and the needs of your students.

How Many Ears Are Here?

Everybody has two ears, but how many ears does the whole class have? Begin a lesson with an intriguing question like this to promote a spirit of investigation in math. And, using their own bodies as a reference is always a motivator for children.

- Ask how many ears one person has. Write the response on a chart.

- Form groups of three or four students. Ask students in each group to determine how many ears their group has altogether. Write this total on the chart.

- Now, ask if anyone can figure out how many ears the whole class has. Students may do this by adding the group totals, using repeated addition of twos, or multiplying. Share solutions on the board. Record all results on a class chart.

Interactive Morning Message:
Today's Factor

Start the day with a Morning Multiplication Message.

Fill a paper bag with number cards from 0 to 10. Select a student to pick out the day's number. Have everyone write down as many multiplication facts as they can that include that number as a factor. For example, if 7 is the day's number, students might write 7 x 3 = 21 or 7 x 7 = 49. This simple exercise accommodates a wide range of ability levels. For example, students looking for a challenge might suggest equations like 7 x 70 = 490 or 35 x 7 = 245.

TIP

Once you solve the "ears question," continue the chart with other body parts, including noses, eyes, and fingers.

Multiplication Secrets
Easy Organizer

So many facts! Many students begin a study of multiplication thinking that the task is huge. The Multiplication Secrets Easy Organizer helps them see that there really aren't many facts that they don't know.

Display a copy of page 86 on an overhead projector. Ask students to count how many of the facts use 0 or 1 as a factor. They might be surprised to find that 36 of the 100 facts use 0 or 1 as a factor. That's more than one third of the list! Once you begin to share ideas like this with students, it doesn't seem so daunting a task. With this in mind, give students a copy of the organizer (see page 78), on which they'll find 100 facts along with simple explanations of four strategies (or secrets). The facts are graphically coded to show which strategy applies to each fact.

For a twist on the flash-card approach to practicing math facts, see *Times Table Mini-Books & Lift-n-Look Flash Cards* by Mary Beth Spann (Scholastic Professional Books, 2000). The Lift-n-Look Flash Cards make easy and portable self-testers that students can use for a quick review. They'll love the quirky cartoons that unfold as they open the flaps to reveal the answers on each flash card.

Literature LINK

Amanda Bean's Amazing Dream: A Mathematical Story

by Cindy Neuschwander (Scholastic, 1998)

Amanda is not convinced multiplication will be much good to her until she has an amazing dream. Use this picture book to introduce the next activity, which helps students make real-life connections to multiplication.

Wheels on a Car

With a little encouragement, students will see that multiplication is all around them—in sets of wheels on a car, in eggs in a carton, and so on.

⑤ Display a model car. Ask students to describe the car using numbers—for example, "A car has two headlights." Have them name other sets of features—for example, four tires, one steering wheel, four doors, two license plates.

⑤ Give each student a copy of page 79. Solve the word problem about the car together. For homework, ask children to draw (or glue) pictures of things that come in sets or have features that come in sets (like the car), then write a related word problem and solution in the spaces provided.

⑤ Have students share their completed papers with the class, then invite them to suggest related multiplication problems. For example, if a student drew a bird with two wings, you could ask, "How many wings would four birds have?" A good culminating activity is to create a class chart of real-life sets that correspond with a given family of multiplication facts—for example: 1 tail on a dog, 2 wings on a bird, 3 wheels on a tricycle, 4 tires on a car, 5 fingers on a hand, 6 sodas in a six-pack, 7 days in a week, 8 legs on an octopus, 9 innings in a baseball game, 10 toes on your feet.

ENGLISH Language LEARNERS

It's always a good idea to give students practice with multiplication in a context, such as in a word or story problem or project format. However, be aware that for English language learners the words may get in the way of the math. To help with this situation, before giving multiplication word problems, create a list of vocabulary that might need to be explained—for example, *product* and *factor*. Help students create and display a chart-paper dictionary with picture or diagram cues for these words. Then look at problems for specific words. For example, a common context for math measurement and multiplication is the *quilt*, a word that may not be familiar to many English language learners. Make sure you identify and define this vocabulary ahead of time. This procedure does double duty. Not only does it ensure that children get to show their math knowledge, it will also help them increase their vocabulary.

Marvelous Multiplying Martians

Children can understand multiplicative relationships long before they begin writing about them. Story and word problems are a natural introduction to learning about multiplication.

Marvelous Multiplying Martians

Early in the morning, (number) Martians came. They multiplied by (number), and played a football game.

(Pause as students use their manipulatives.)

⑤ Supply each student with a set of 20 manipulatives such as counters, beans, or cubes.

⑤ Read the poem "Marvelous Multiplying Martians." (See right.) As you read the poem, have students model the action with their manipulatives and find the answer to the final question in the poem. Model this the first time around, showing how the first number given represents the number of Martians in a set and the second number represents the number of sets. So, if we have three Martians landing in the first set, and then we say they multiplied by four (meaning there are now four sets), the students make four sets of three, count, and find the answer.

When earthlings saw them playing, the whole crowd shouted, "Wow!" Just how many Martians, do you think, are playing football now?

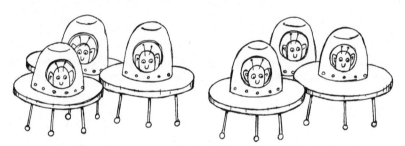

Literature LINK

2 x 2 = BOO! A Set of Spooky Multiplication Stories

by Loreen Leedy (Holiday House, 1995)

Witches, cats, bats, skeletons, mad scientists, and other spooky creatures get involved in situations that use multiplication from zero to five in this colorful, cartoon-filled picture book. There are plenty of puns and gruesome goings-ons to get kids chuckling and thinking about multiplication.

Potato Print Sets

Pictorially creating and then counting sets is a good way for students to explore multiplication ideas.

- Provide printing ink or tempera paint and art paper, and give each student scissors and half a potato. Help students carve out a shape on the flat side of the potato with the scissors. A simple shape like a triangle or square is best.

- Have students use a paintbrush to coat the raised shape with ink or paint, then use the potato to print the shape repeatedly on the paper. Allow students to print freely in patterns and sets of their own design the first time around.

- Ask students to print again on a second sheet of paper, but this time to arrange their shapes in a set. Model this on the board, drawing a set of three stars and circling them to indicate it is a set. Have students stamp their own set of a number of objects, then repeat the set five times.

- When they are finished, ask them to find out how many of their designs are on the paper.

- Discuss methods and strategies they used, then share how multiplication (counting the number of sets and multiplying that by the number of objects in the set) can give you a quick and accurate answer.

Cookie Sheet Math

Arrays are a great way for kids to see a pictorial or physical representation of multiplication.

Explain to students that an array is an arrangement of objects in rows and columns. To explore arrays, give each student a box of animal crackers and a napkin. Have students open up the napkin, then arrange 12 cookies in rows and columns—for example, three rows of 4 or two rows of 6 or four rows of 3 or one row of 12. Invite students to take an "idea walk" around the room to see what other sorts of arrays there are. Use a variety of target numbers for students to explore. At the end, use subtraction to remove all the cookies. (That's right—eat them!)

Sue Lorey
Grove Avenue School
Barrington, Illinois

TIP

If you don't have time to incorporate the printmaking aspect into your lesson, rubber stamps and ink pads also work well.

TIP

Check for food allergies before letting children eat the animal crackers.

Array for Math Display

Use arrays to create a bulletin board reference that helps students conceptualize what multiplication facts are really about.

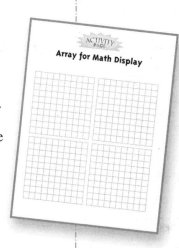

Assign a set of multiplication facts to a group of students. For example, one group might be assigned the fours. Give children copies of the centimeter squares (see page 80) and have them cut out arrays that match each assigned fact. Arrange the arrays on a bulletin board. Have children record the facts that match their arrays and add those to the display. When they're finished displaying arrays and facts, students will have a complete multiplication table reference and will have had a very meaningful math experience.

Marie Mastro
Dutch Neck School
Princeton Junction, New Jersey

Sticky-Dot Arrays

Here's a tip for helping children make arrays quickly and easily.

Most office supply stores carry sticky dots in many colors, including everyone's favorite—fluorescent! These self-stick dots are perfect for creating arrays—all students have to do is peel and stick. Give each student a sheet of drawing paper and a set of sticky dots. Then deal out multiplication cards from a deck, and have students create an array for the card they get. When everyone is finished, let students take turns holding up their arrays while the rest of the group tries to identify the multiplication fact and answer that matches the array.

Susan Coleridge
Wicoff School
Plainsboro, New Jersey

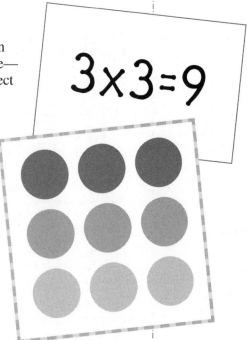

Clap Them Out!

Add a little rhythm to multiplication facts practice to engage more of the ways your students learn.

Begin by writing a multiplication sentence on the board—for example, 4 x 4. Tell students you will be clapping together to find the answer. The first factor normally tells us how many sets of objects there are, and the second factor tells how many objects are in the set. So, we would clap out this problem like this:

> **Clap, clap, clap, clap (pause)**
> **Clap, clap, clap, clap (pause)**
> **Clap, clap, clap, clap (pause)**
> **Clap, clap, clap, clap.**

Students can clap along with you and even count aloud as they go. When you've finished clapping, have everyone call out the answer together. Let students take turns recommending facts to try and lead the clapping. Finally, try clapping out a few facts without identifying the multiplication sentence first. Challenge students to listen carefully enough to identify the fact and the product you are clapping.

Soda Six-Pack Demonstration

What is multiplication really about? This demonstration helps children understand it's about answering the simple question, "How many?"

⑤ Bring in five six-packs of soda or juice and place them on a table. Gather the class and ask, "How many cans are here?" Give students a minute or two to think about the problem, then ask for solutions and strategies. Mention that you could count them by ones and demonstrate. (Ask a student to time you with a clock or watch.)

⑤ Now, point out that you could also repeatedly add sixes (five times). Do this as another student times you.

⑤ Finally, ask if you could use multiplication. What multiplication fact would describe this situation? 5 x 6! Have a student time you as you say, "5 times 6 is 30." Ask: "Of the methods we tried, which do you think was the fastest and most accurate?" *(knowing the multiplication fact)* That's why we have multiplication facts and work toward memorizing them—they give us a very efficient way for finding out "How many?"

TIP

Let students work in groups to practice presenting information. Assign a strategy, or have students suggest a strategy, for each group. Have students discuss the strategy and then prepare a presentation that includes an explanation, a diagram or picture to illustrate an example, and a list of facts the strategy applies to.

Find All the Facts!

Here's a simple but effective way to help children discover various multiplication facts that create a given product.

You'll need a set of small candies, such as M&Ms® (or raisins, O-shaped cereal pieces, beans, or counters), and an overhead projector. Let's say the product is 12. Provide each student with 12 candies. Put 12 on the overhead as well. Ask students to arrange their 12 candies so that they form an array. Ask what multiplication fact describes the array. Now ask them to rearrange the M&Ms to form a different array. The product is still twelve, but what's the new multiplication fact? Investigate possible facts and record them on the bottom of the transparency. Students can record facts on their papers. Work through products from 1 to 100 and find all the facts!

Literature LINK

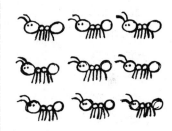

One Hundred Hungry Ants

by Elinor J. Pinczes (Houghton Mifflin, 1993)

One little ant leads the rest of the group to the food. They march in a variety of formations, which are actually different arrays. Students will enjoy reading along and describing or writing the multiplication sentences that match the arrays the ants form as you read. This is an especially suitable read-aloud for the 100th day of school!

Ants in Arrays

For even more fun and learning, you might want to create a bulletin board display based on *One Hundred Hungry Ants*.

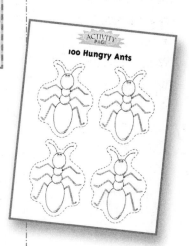

Give each student a copy of page 81. If you have 25 students and each one colors and cuts out the four ants on the reproducible, you'll have 100 ants. Now, look at the bulletin board space with students. What arrangement or array of ants would fit best? Test a few out on the floor, and describe them in terms of multiplication—for example, 5 rows of 20 or 5 x 20. Decide together on the best way and complete the bulletin board together.

Skip Along the Hundred Chart

One of the best beginning bridges to understanding multiplication is skip counting. Use the Hundred Chart on page 82 to help students practice this skill.

Make several copies of page 82 for each child. Begin the lesson by having students color in the numbers on their hundred chart as they count by twos. (Students count and color 2, 4, 6, 8, 10, 12, and so on.) Guide children to notice how this pattern relates to multiplication facts with twos. Have students look at the products on a twos times table and compare this to the hundred chart pattern. They are the same numbers! Do this for each multiplication table as you introduce it: threes, fours, fives, and so on. Students can record the facts on the back of their hundred chart and compare the patterns of products. Collecting these pages in a book provides a handy reference.

Wendy Weiner
Parkview School
Milwaukee, Wisconsin

Multiplication Necklace Match

Not only does this activity help students model facts, but it is also a good beginning-of-the-year, getting-to-know-you opportunity.

Give each student yarn and scissors, as well as beads in two colors. Also provide a card with a multiplication fact on it. The task is to create a necklace that models that multiplication fact. For example, if the fact is 3 x 6, the student would need to use 18 beads. To model three sets of six, he or she might then string a set of six black beads, then a set of six white beads, followed by another set of six black beads. Once the beads are strung, students can cut the yarn and tie the ends so that the necklace fits easily over their head. When everyone is finished, collect the cards and shuffle them. Pass them out randomly to students and have them find the classmate who is wearing the matching necklace.

Multi-Link Multiplication Buildings

Children love building with multi-link cubes. Tap into this motivation with a multiplication building activity.

Give each student a set of at least 20 cubes. Explain to students that each cube is a room in a building they are going to make. Tell the class, "You will begin by making the first floor of the building. Each floor will have three rooms (or cubes). Link three rooms (cubes) together to make the first floor of the building." Model this and ask students to make their own, in any configuration they like. Have students add a second floor to their buildings, making it an exact duplicate of the first floor. Ask students to determine how many cubes they've used so far. Have them share how they arrived at their answers, then give a multiplication sentence that describes the building so far (2 x 3). Record this on the board and have a student explain how it matches the building. Continue adding floors and recording multiplication sentences that match. Change the first floor to two cubes, four cubes, or any other factor you want the class to work with.

Adapted from Investigations in Number, Data, and Space: Grade 3 *by TERC (Dale Seymour Publications, 1998)*

Multiplication Mix-Up

Have students really experience multiplication vocabulary by "becoming" the number sentence.

- Make oaktag "signs" for the following: the word *factor* (make two), the word *product*, (make one), an equals symbol (one), a multiplication symbol (one), the numbers 0 to 100 (one number per card).

- Choose five children, and have them stand shoulder to shoulder in a line. As a demonstration, pick up three number signs—for example, 3, 6, and 18. Give the first child a "3" sign and a "factor" sign. Give the second child the "multiplication symbol" sign. Give the third child a "6" sign and a "factor" sign. Give the fourth child the "equals symbol" sign, and the fifth child the "18" sign and the "product" sign. A correct number sentence has been formed!

- Repeat, giving new groups of children cards, and having them arrange themselves to form a correct number sentence.

Turn the activity into an exciting game format. Put vocabulary, symbol, and selected number signs in a pile. Choose five students. Tell them they have 30 seconds to find the correct signs, put themselves in order, and create a correct multiplication sentence.

Hundred Chart Exploration

When we look at the times tables from zero through nine or ten, there are a lot of products covered. Is there a basic (one digit times one digit) multiplication fact that yields a product for every number on the hundred chart?

Give students a chance to explore this question with a copy of the hundred chart (see page 82) and a pencil or crayon. Ask students to look over the chart and color in any numbers that could be products for basic multiplication facts. Ask them to record at least one of the facts that gives that number as a product. When they're done, there will be plenty of numbers not colored in. Ask students why they think this is so. What do those numbers have in common? Are there any patterns? This is an interesting exploration that helps students examine number patterns, while again practicing facts.

To take Mr. X's clothing combinations further, give students additional copies of page 83. Let them make more colored shirts and pants and test out whether the multiplication and pattern will work with even higher numbers.

Dress Mr. X

Help students discover the relationship between combinations and multiplication with the mysterious Mr. X and his manipulable wardrobe.

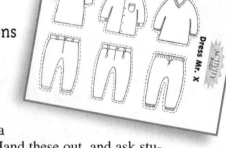

- Prepare for the activity by making a copy of page 83 for each student. Hand these out, and ask students to color the shirts and pants so one of each is blue, one is green, and one is yellow, then cut them out.

- Explain that this is the wardrobe for Mr. X. He wants to know how many different outfits are possible by mixing and matching his shirts and pants. Ask students to use the manipulatives to find all the possible combinations. Do a couple together for practice: Starting with one shirt and two pairs of pants, how many combinations are possible? It might be worked out like this:

> **Blue Shirt - Green Pants**
> **Blue Shirt - Yellow Pants**
> **Two combinations are possible (1 x 2 = 2)**

If you add another shirt, things change:
Blue Shirt – Green Pants
Blue Shirt – Yellow Pants
Yellow Shirt – Green Pants
Yellow Shirt – Yellow Pants
Four combinations are possible (2 x 2 = 4)

After trying two of these together, ask students to work out the others on their own. If students work out enough of these combinations, they will begin to see the pattern and discover that multiplying the number of pants by the number of shirts will give them the number of possible combinations.

Multiplication Golf Board Game

Warning! This game is more addictive than real golf and much cheaper. Don't be surprised if you find yourself looking for opportunities to play.

- Ask students if anyone has ever played golf or knows how it is played. Have students share their ideas. Make sure everyone understands that there are 18 holes and that the golfer tries to get the ball in each hole. He or she tries to do this with the fewest number of swings (or strokes) possible. The person with the lowest number of swings at the end of the 18 holes is the winner.

- Pair up students for their own game of golf. Give each pair a copy of pages 84 (the game board) and 85 (the score card) and two dice.

- Students take turns moving to each "hole" in order, one at a time. When they get to a hole they roll the dice, multiply the numbers, and compare the product to the requirement at that hole. For example, if the roll is 3 x 3 (9), and the requirement is that the product be an odd number, they've put the ball in the hole and used only one stroke (or roll). If the product does not meet the criterion of that hole, the player must record that stroke and roll again. That's a second stroke. Each time a player rolls, another stroke is recorded. At the end of the 18 holes, the player with the fewest number of strokes (rolls) is the winner.

Sue Lorey
Grove Avenue School
Barrington, Illinois

Multiplication Toss

Here's an easy (but very motivating) multiplication game to make and play.

Ask students to bring in egg cartons. Have them open the top of the egg carton, and in the 12 spaces, mark the numbers 0 through 11 in any order they wish. Provide each student with two table tennis balls (or other small objects such as beans or cubes). Have students pick a partner and put their egg cartons side by side on the floor. Students stand a couple of feet away from their cartons and take turns tossing the two objects into their cartons. The spaces the objects land in become the factors for a multiplication problem. Players do the multiplication and then compare products. The highest product wins!

 ## Multiplication Mamba

As students begin to practice multiplication facts, rhythm and song can make the whole process more fun and effective.

Use a small drum as you sing and beat out the rhythm of this silly song. Have students clap and sing along, or provide them with rhythm instruments to play. Here's what 8 x 4 = 32 would be like. Chant or sing:

> **Mamba, Mamba, da, da,**
> **eight times four is . . .**
> **Mamba, Mamba,**
> **thirty-two.**

The second line of "mamba, mamba" provides students with a few seconds to recall the answer to the fact. Chant through the times tables, first reading them aloud from a chart when chanting, then working from memory.

Judy Wetzel
Woodburn School
Falls Church, Virginia

For more fun, try *Tunes That Teach Multiplication Facts* by Marcia Miller with Martin Lee (Scholastic Teaching Resources, 2005).

Multiplication War

Playing cards are always a reliable and motivating math manipulative.

- Use any deck of playing cards to play this version of War. Kings and Queens are worth ten. Jacks are zero. Aces are one.

- After pairs of players deal the cards (evenly between them), they each place the top two cards faceup and multiply the number values of the cards.

- Players compare products, and the player with the highest product wins the cards in that round. If players tie (products are the same), they each place three cards facedown, and two more cards faceup. They multiply the last two cards, then compare products.

- After players are out of cards (or don't have enough for another round), they count up the cards they've won to see who has the most.

School Store

A school store is a great long-term project to promote multiplication practice.

Many inexpensive school-supply items, such as erasers and pencils, can be purchased at party stores, dollar stores, or through catalogs. Let students help order items and set up, organize, and price inventory. All you need is a desk or table in the hallway and about 20 dollars in change to start with. Let the rest of the school know about the store. Set up a schedule, assigning certain dates and times when other classes can come shopping. Schedule your class to serve as clerks on a rotating basis. Since most items are inexpensive, it is very common for purchasers to buy multiples of erasers, pencils, and other small items. Your class will find plenty of real opportunities for practice with money, time, scheduling, addition, subtraction, making change, and, of course, multiplication.

Fact Riddles

Children naturally enjoy telling and making up riddles. You can give them plenty of opportunities, even in math, when they create "Fact Riddles."

Give each student a multiplication fact card. Challenge everyone to write a riddle or set of clues that tells about the three parts of the number sentence. Have students write each clue on a separate index card. For example, for the fact card 7 x 3 = 21, the first clue might be: "I am the age of many first graders." The second clue might be: "I am the number of wheels on a tricycle." The third clue might be: "If you add my two digits together they equal three." When a student takes on the challenge to guess the fact, he or she can ask for clues one at a time in any order—for example, starting with the product clue. As long as they keep getting the correct answers to the clues, they can keep going. If they miss, the guessing goes on to another person. Once all three parts are guessed and the entire problem is revealed, the next presenter takes a turn.

For more fun with math riddles, try *Mighty Fun Multiplication* by Bob Hugel (Scholastic Professional Books, 2001). When children solve the math problems on each page, they get the answer to a rib-tickling riddle!

Literature LINK

Sea Squares

by Joy N. Hulme (Hyperion, 1991)

This a lovely picture book for introducing students to multiplication. The rhyming text and beautiful ocean-themed illustrations really appeal to children and hold their interest. Each page features a sea creature and natural multiples described in upbeat rhyme, such as "Four slippery seals, with four flippers each, swooping in the surf and flopping on the beach. Four seals are quite complete, with 16 flippery feet." Children will enjoy counting and verifying the text. As a further challenge, ask students to write the multiplication sentences that match the pictures and text.

Easy and Hard Sort

Help students overcome the idea that "There are SO many facts!" with this activity.

○ Give each child scissors, a copy of page 86, two letter-size envelopes, and a two-pocket folder. Tell students that you know it seems that there are a lot of facts out there, but actually they already know most of them.

○ Explain to students that they are going to do an "Easy and Hard Sort." Call out some simple facts like 1 x 1 or 2 x 2. Ask: "Are these easy or hard?" Ask students to cut apart the facts, then place the ones they know automatically in the envelope marked Easy. Have them put the facts that are still a challenge in the envelope marked Hard.

○ Have students compare the contents of the two envelopes. There will surely be more in the Easy envelope. Now, point out to students that they have self-assessed and now know which facts they need to practice the most. As students learn strategies and practice their facts, they'll move more from the Hard envelope to the Easy one.

○ During independent work time, students can take facts from their Easy envelope and glue them to the inside of their folder so their fact mastery becomes a more visible reminder of their progress.

Multiplying Mummies

Tell students it's time to visit ancient Egypt and see how mummies like multiplication.

○ Give each child a sheet of paper, a pencil, and a pair of dice.

○ Have students roll one die. Whatever number comes up, that's how many pyramids (triangles) they draw on their paper.

○ Have students roll the second die. That's how many mummies they put in each pyramid. (Students can represent mummies by writing the letter M or drawing any simple symbol.)

○ Ask students to write a number sentence that shows the number of pyramids times the number of mummies. The total will tell how many mummies are in all the pyramids.

To help students begin to get an overall visual picture of the 100 multiplication facts, give them a copy of the graphic organizer on page 87. This displays the facts in a grid and features a format where product patterns are visible as well as the pattern of equations. It's an easy reference to work from when looking for patterns and answers or as a tool to aid in fact memorization.

Enough for Everyone: A Party Project

As adults, we commonly use multiplication when planning dinners and shopping. Let students plan a party and multiply as needed.

Before photocopying a class set of the party planner on page 88, make a master for yourself and fill in the prices that you want students to work with. In this way, you can individualize the activity to meet the level of complexity you want for your group. Give each child a copy of the planner. Have children choose items from the advertisement at the top of the page, then complete the order form and calculate the total cost. Answers will vary as students make different choices about what to have at the party. After practice on this page, give it a try with the real thing—figure out an order as a class—then have that party!

Literature LINK

Just Add Fun!

by Joanne Rocklin (Scholastic, 1999)

In this short, beginning chapter book from the Hello Math Reader series, two boys plan a party and use multiplication to figure out how much food they need, how they can pay for it, and how to get everyone seated. The last section of the book includes excellent activities by Marilyn Burns, noted math educator, on learning multiplication with skip counting, repeated addition, and arrays.

Mysterious Multiplier Posters

Children work with partners to create mysterious multiplication problems for their classmates to solve.

Assign a pair of students a set of facts (twos, threes, fours, etc.) and ask them to keep this information a secret. Have students write out the facts for that set as follows: write the first factor, leave out the function sign and the second factor, and then write the equals sign and the product. Have them write out the full set, but not necessarily in order. The threes might look like this:

$$7 \ \underline{\hspace{1.5em}} \ \underline{\hspace{1.5em}} = 21$$
$$9 \ \underline{\hspace{1.5em}} \ \underline{\hspace{1.5em}} = 27$$
$$2 \ \underline{\hspace{1.5em}} \ \underline{\hspace{1.5em}} = 6$$

Give the partners a sheet of black construction paper, and have them go to the front of the room with their fact sheet. Have one partner slide the black paper down to reveal the first fact, then the second, then the third. The first person to guess the rule goes next. For an even more challenging version, mix functions and fact families on the same poster.

Multiplication Improvisation

Comedy improvisation shows are very popular. Often a comedian is handed a slip of paper with a word or phrase on it, and he or she must quickly integrate that idea into the act. Improvisation can add a good deal of fun to multiplication practice, too.

- On slips of paper, write some simple story starters such as, "One day I was visiting the zoo . . ." or "It was a dark and stormy night . . . ." Place them in a box. Write some multiplication problems on slips of paper and place them in another box. Write nouns such as *dogs, clocks, spoons,* and *monkeys* on slips of paper and place them in a third box.

- Let students take turns choosing a slip from each box, then use them to create a multiplication story that incorporates all three elements and mentions the product. A student who picked "One day I was visiting the zoo . . . ," the multiplication sentence "2 x 9," and "monkeys" might create this multiplication story: One day I was visiting the zoo when two sets of nine monkeys escaped. I chased them everywhere but could not catch all 18 of them.

Keep the improvisation moving along by giving students 30 seconds per multiplication story.

Kid-Created Concentration

This is a good activity for a center that an individual or a small group can enjoy independently.

○ Stock a center with white index cards and a marker. Have students use the materials to create two sets of cards. On one set students write multiplication sentences, such as 5 x 5; on another set of cards they write the product, such as 25. They do this for all the facts you want them to practice. A reasonable number of facts would be 20—perhaps the eights and nines tables.

○ Have students put the two sets of cards together, mix them up, then place them facedown in five rows of four. Have players take turns turning up two cards at a time and trying to achieve a match of a number sentence and its correct product.

Roll a Product

This simple game generates countless facts to practice in a fun format.

Provide pairs of students with two dice or number cubes and a copy of the Roll a Product Score Sheet. (See page 89.) Have students take turns rolling the dice. They write whatever numbers come up in the factor boxes of their equation blanks, then multiply the two numbers to get the product. Whoever gets the higher product wins that round. For a change of pace, have the smaller product win.

Jacqueline Clarke
Cicero Elementary School
Cicero, New York

Take-Home Activity:
Five in a Row

Keep homework fun but challenging with assignments in game formats. This one lets children practice multiplication facts while they're enjoying a game with a family member.

○ Make a copy of the Five in a Row game board (page 86), the Product Cards (page 90), and the directions (page 91). It is best to copy these on a sturdy paper like oaktag (or copy on regular paper and then glue to oaktag) and laminate.

○ Cut up the product cards and put them in a reclosable bag. Put this bag, along with the game board and directions, in a larger bag to create a Take-Home Math Pack. Review the game with students. (You might also play a round in class to demonstrate.) Let children sign out the math pack to play at home.

ENGLISH Language LEARNERS

All students benefit when they are given an opportunity to explain their thinking in math, but this is especially true of English language learners. When we acquire another language, our receptive language becomes stronger first—that is, we understand the language when we hear it. Our productive language, or speaking of the language, comes later. To help support the development of productive language, be sure to provide practice in explaining thinking in math. A simple way to do this is to pair up students and give each a different multiplication word problem. Ask students to solve their problems and then show their answers and work to their partner. Have them explain how they did the problem and why they think their answer is correct. Be sure each student has a chance to take a turn with this. This is good practice for everyone, and doing it in the small, controlled environment of a pair/share, rather than in front of the whole class, enables students to gain confidence with both language and mathematics skills.

Multiplication Story Book

It's always fun to create a class book. A Multiplication Story Book is no exception.

Assign each student a page in the book to complete. Have students write and illustrate a short story problem that could use multiplication to find the answer, then use the back of the paper to show how to solve the problem. Have students do a rough draft first and edit with a partner for both writing and math accuracy. Collect and bind the pages to make a class book, or photocopy so each student has a copy to work through.

Slow Facts, Fast Facts

When helping students learn the times tables, include as many of the senses as possible. Here is an activity that simultaneously engages touch, sight, and hearing for maximum sensory input.

- Start by chanting a table together, such as "5 times 1 is 5, 5 times 2 is 10, 5 times 3 is 15, 5 times 4 is 20."
- Clap on each syllable, emphasizing the last syllable.
- Play around with the rhythm, slowing it down and speeding it up.
- As you clap, have a multiplication times table chart available and a couple of cheerleaders pointing out the facts and the answers.
- Later, hide the answers, but keep the cheerleaders pointing to the facts as they go. In this way, students put touch, sight, and hearing together to practice their facts.

Take-Home Activity:
The Great Multiplication Scramble

This activity lets children and families have fun working together to practice multiplication facts.

Give each child a copy of page 92. Various numbers are scattered all over this page. The object of the Scramble is to try to find as many complete multiplication number sentences as possible. Assign the page as homework, asking children and families to work together to find the maximum number of multiplication sentences that they can, and record them at the bottom of the page. (Remind students to share the directions with their families before beginning.)

Math Walk

Kids are always on the move. How far do they walk in a typical day or even in an hour? Here's a challenging and engaging project that links physical education and mathematics in a very real way. It also provides a natural opportunity for multiplication on a calculator. You'll need several pedometers. (Check teacher supply catalogs.)

🌀 Have students measure the length of each other's stride. To do this, have them walk a pre-measured 50-foot distance while counting their strides. They then divide the 50 feet by the number of strides and arrive at their stride length.

🌀 Introduce the pedometer. Explain that this instrument will record the number of steps a student takes in a given period of time. Invite children to share their explanations for how this can help them know the total distance they walk. (Multiplying the student's stride measurement by the number of strides equals the total distance walked.) Have students take turns with the pedometers to calculate how far they walk during the day. For example, they might calculate the distance between the classroom and other places they go, then determine a total for the day.

Lynn Holman
Upper Elementary School
Plainsboro, New Jersey

Let four funny superheros in the software program *Mighty Math Calculating Crew* (Edmark, 1996) entertain students while reinforcing math skills.

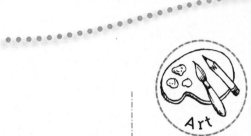

Doubles Poetry Garden

Art

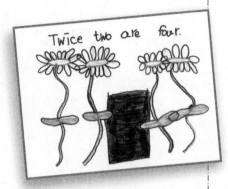

One of the best strategies for learning multiplication facts is to learn the doubles. The poem on page 93 from *One, Two, Skip a Few: First Number Rhymes* (Barefoot Books, 1999) is a great way to help children do just that.

- Give each child a copy of the poem (page 93). Read the poem aloud.

- Assign stanzas to partners or small groups. Display a mural-size sheet of paper and give each group a section.

- Have students write their verse on the paper and write the corresponding multiplication fact underneath. Provide children with field guides on flower identification. (See tip, left.) Have them use the reference material to illustrate their verses. When completed, the mural makes a great artistic multiplication reference!

Wendy Wise-Borg
Rider University
Lawrenceville, New Jersey

TIP

Help young readers identify hundreds of flowers found in North America with the photos and easy-to-read text in *The National Audubon Society First Field Guide: Wildflowers* (Scholastic, 1998).

Literature LINK

The Grapes of Math

by Greg Tang (Scholastic Press, 2001)

Clever rhymes matched with colorful illustrations of groups of objects like bees, fish, cherries, and camels make for an interesting excursion into poetry and math. Each rhyme describes the set of objects in the picture and a way to count them. The reader must use the clues in the rhyme to determine the number of items. The counting strategies hinted at in the clues include a variety of operations and problem-solving strategies and are described in detail in an index in the back of the book.

Multiplication for Estimation

Estimation is an important skill to develop in all areas of mathematics, and the more strategies students have for estimating, the better their estimations will be. Multiplication is a helpful strategy to use in estimation.

- Display a container filled with a large number of a small item such as gum balls or buttons. Ask students to estimate how many of the item are in the container. Record, compare, and discuss the estimates.

- Now, remove a handful of the items and count them out. Let's say there were 20 gum balls in your handful. Ask students if knowing this will help them in revising or confirming their estimates. A good strategy here is to then estimate how many handfuls would fit in the container. A student who estimates that about ten handfuls are in the container can then multiply 10 x 20 and conclude that an estimate of 200 gum balls would be reasonable.

- Practice this to help students begin to "eyeball" what is in the container by estimating how many of the item might fit in a layer or handful and then multiplying by their estimated number of layers or handfuls. This helps students bring reasoning into the process of estimation and thus create more accurate estimates.

Gail Englert
Sewells Point Elementary
Norfolk, Virginia

Anno's Mysterious Multiplying Jar

by Masaichiro and Mitsumasa Anno (Philomel, 1983)

Highly detailed drawings help tell the story of a mysterious jar and what's inside. This picture book will extend your students' multiplication thinking in some complex ways!

One popular way to assess multiplication fact knowledge is through timed tests. This is fine. Automatic recall of facts is an appropriate goal for third graders. However, using these too often can result in more problems than learning. Use timed fact tests at intervals to track progress once a month, or once each marking period. Have students practice facts through lessons, activities, exploration, and games.

Multiplication Magic

Children love a good magic trick. Here's one that uses the sevens times table and is a snap to learn.

◎ Have a student throw any number of dice. Can anyone tell the total of the top and bottom numbers (without looking at the bottom)? Read ahead to learn the trick, then stun them with your magical powers by telling them the answer (the more quickly, the better). Try it again a few times to see if anyone catches on. The secret? On any die, the top number plus the bottom number equals seven. So if four dice are thrown, the top and bottom numbers will add up to 28 every time. Five dice will always yield 35. Six will give you 42. And so on.

◎ Provide plenty of dice and let children take turns practicing the trick with one another.

◎ Then let them take their magic show on the road to wow other classes. They'll have fun telling their audience that they are able to see through the dice with their X-ray vision and that this, coupled with their incredibly fast calculating ability, enables them to find the answer almost instantly!

Adapted from Mathemagic *by Raymond Blum (Sterling, 1991)*

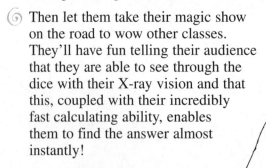

TIP

Advise students to start with two dice and work their way up to more as the audience becomes increasingly amazed.

Literature LINK

Can You Count to a Googol?

by Robert Wells (Albert Whitman, 2000)

Wells uses lots of multiplication and a cartoon illustration style to bring his readers along through millions, billions, trillions, and the googol! (It has a hundred zeros!)

Test-Taking Tip

A common mistake among students on tests featuring computation is to add instead of multiply. Teach students how to avoid this mistake with this test-taking strategy.

Explain to students that on multiple-choice tests, one of the answer choices for multiplication sentences is usually the sum of the two numbers. This is called a distractor. Share examples of distractors with students. Advise students to always check their signs before choosing an answer. The best way to practice this test-taking strategy is with pretend tests that include distractors. Ask students to circle the distractor in red crayon and the correct answer in pencil. By making students aware of distractors, they are less likely to be confused by them during real test situations.

Take-Home Activity:
Fill in the Missing Product

Children and families have fun practicing multiplication with a puzzle-like activity.

Make a copy of page 94 for each student to take home. Ask children to work with a family member to complete the table, just as they might work with a friend to complete a puzzle. It's fun to find all the patterns and feel the satisfaction of completing the table. The completed paper becomes a great at-home reference for practicing multiplication and learning times table patterns.

Target Number

Play this game for a quick, fun way to practice multiplication.

⊚ Begin by telling students, "I'm going to give you the answer. Here it is: 30!"

⊚ Have students write as many different multiplication facts as they can that would give them that product. The more the better!

Go to **www.aplus math.com** for individualized online fact practice, worksheets, puzzles, games, a homework helper section, and a flash-card creator. This is also a good site to recommend to families with Internet access as a way to help students practice multiplication facts at home.

Name _____   Date _____

Multiplication Secrets
Easy Organizer

	0	1	2	3	4	5	6	7	8	9
0	0	0	0	0	0	0	0	0	0	0
1	0	1	2	3	4	5	6	7	8	9
2	0	2	4	6	8	10	12	14	16	18
3	0	3	6	9	12	15	18	21	24	27
4	0	4	8	12	16	20	24	28	32	36
5	0	5	10	15	20	25	30	35	40	45
6	0	6	12	18	24	30	36	42	48	54
7	0	7	14	21	28	35	42	49	56	63
8	0	8	16	24	32	40	48	56	64	72
9	0	9	18	27	36	45	54	63	72	81

Zappy Zeroes: Any number times zero equals zero. (19 facts)

Wonderful Ones: Any number times one equals the original number. (17 facts)

Doubles: These are like adding the number to itself. (2 x 6 is like 6 + 6 = 12) (15 facts)

Nifty Nines: Subtract one from the second factor. That is the first digit of the product. The second digit of the product is the number you need to add to the first digit to make 9. (13 facts)

Note: Many facts can be solved with more than one strategy.

The Great Big Idea Book: Math © 2009, Scholastic Teaching Resources

Name _____ Date _____

Wheels on a Car

Look at the picture of the car. Read the word problem about wheels on cars. Write a solution to the problem. Then look around for objects that include other sets. Draw or glue pictures to show what you find. Write a multiplication problem for each set. Solve it!

Draw or glue a picture of a set here.	Write a multiplication problem about the set here.	Write your solution here.
	There are four wheels on a car. How many wheels on three cars?	

Array for Math Display

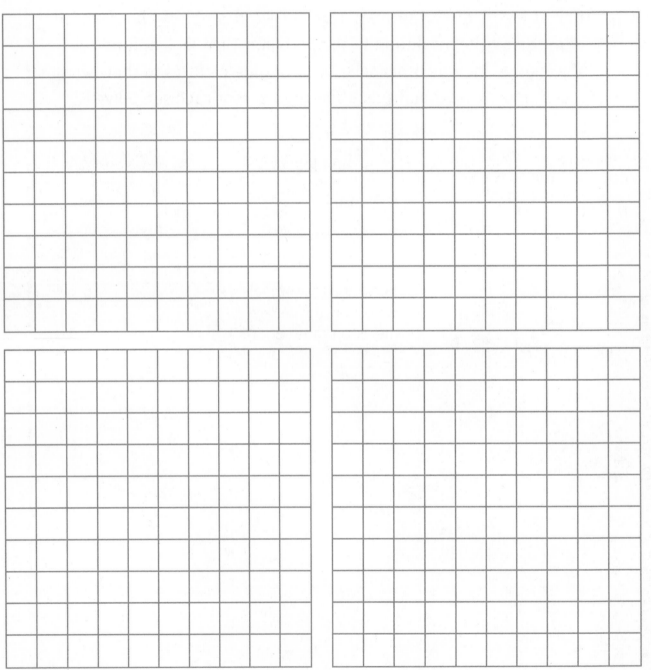

The Great Big Idea Book: Math © 2009, Scholastic Teaching Resources

100 Hungry Ants

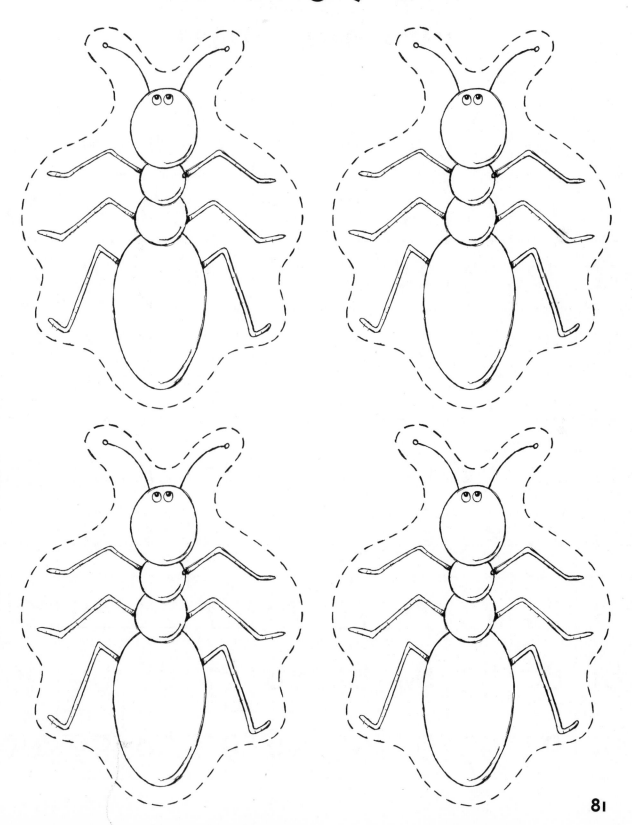

The Great Big Idea Book: Math © 2009, Scholastic Teaching Resources

Name _____   Date _____

Hundred Chart

1	2	3	4	5	6	7	8	9	10
11	12	13	14	15	16	17	18	19	20
21	22	23	24	25	26	27	28	29	30
31	32	33	34	35	36	37	38	39	40
41	42	43	44	45	46	47	48	49	50
51	52	53	54	55	56	57	58	59	60
61	62	63	64	65	66	67	68	69	70
71	72	73	74	75	76	77	78	79	80
81	82	83	84	85	86	87	88	89	90
91	92	93	94	95	96	97	98	99	100

The Great Big Idea Book: Math © 2009, Scholastic Teaching Resources

Dress Mr. X

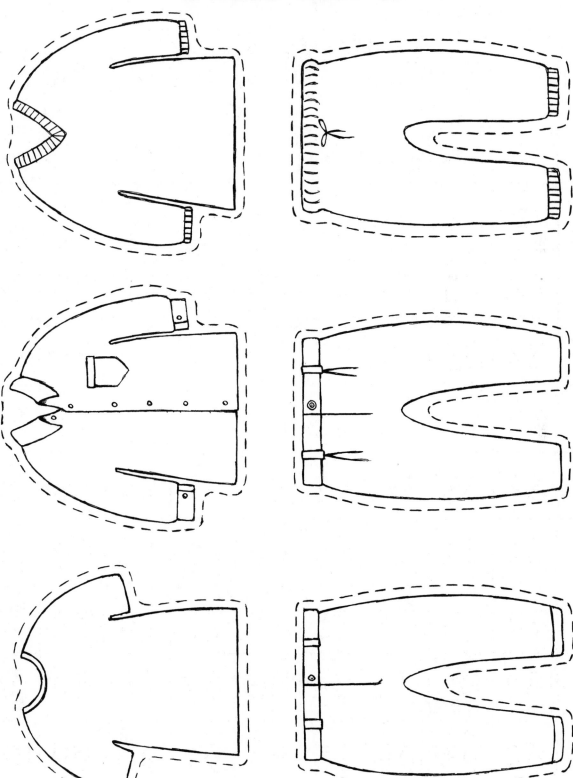

Name _____

Date _____

ACTIVITY PAGE

Multiplication Golf

Holes:
- 18 greater than 9
- 17 even
- 16 less than 18
- 15 odd
- 11 less than 20
- 12 greater than 12
- 10 between 15 and 40
- 14 between 4 and 16
- 13 greater than 7
- 9 odd
- 8 even
- 7 greater than 4
- 6 between 8 and 15
- 5 lower than 10
- 4 lower than 30
- 3 greater than 15
- 2 even
- 1 odd

84

The Great Big Idea Book: Math © 2009, Scholastic Teaching Resources

Multiplication Golf
Score Card

Player's Name _____

Hole	Strokes (Rolls) Tally	Strokes (Rolls) Total	Running Total	Rolls
Example	|||	3	3	2x3=6 4x4=16 3x1=3
1				
2				
3				
4				
5				
6				
7				
8				
9				
10				
11				
12				
13				
14				
15				
16				
17				
18				

Final Score

Remember: In golf, the LOWEST score is the winner!

Name _____   Date _____

× 0 0	× 0 1	× 0 2	× 0 3	× 0 4	× 0 5	× 0 6	× 0 7	× 0 8	× 0 9
× 1 0	× 1 1	× 1 2	× 1 3	× 1 4	× 1 5	× 1 6	× 1 7	× 1 8	× 1 9
× 2 0	× 2 1	× 2 2	× 2 3	× 2 4	× 2 5	× 2 6	× 2 7	× 2 8	× 2 9
× 3 0	× 3 1	× 3 2	× 3 3	× 3 4	× 3 5	× 3 6	× 3 7	× 3 8	× 3 9
× 4 0	× 4 1	× 4 2	× 4 3	× 4 4	× 4 5	× 4 6	× 4 7	× 4 8	× 4 9
× 5 0	× 5 1	× 5 2	× 5 3	× 5 4	× 5 5	× 5 6	× 5 7	× 5 8	× 5 9
× 6 0	× 6 1	× 6 2	× 6 3	× 6 4	× 6 5	× 6 6	× 6 7	× 6 8	× 6 9
× 7 0	× 7 1	× 7 2	× 7 3	× 7 4	× 7 5	× 7 6	× 7 7	× 7 8	× 7 9
× 8 0	× 8 1	× 8 2	× 8 3	× 8 4	× 8 5	× 8 6	× 8 7	× 8 8	× 8 9
× 9 0	× 9 1	× 9 2	× 9 3	× 9 4	× 9 5	× 9 6	× 9 7	× 9 8	× 9 9

The Great Big Idea Book: Math © 2009, Scholastic Teaching Resources

Use with "Easy and Hard Sort" and "Five In A Row"

Name _____   Date _____

Product Patterns
Graphic Organizer

X	0	1	2	3	4	5	6	7	8	9
0	0 0 x 0	0 0 x 1	0 0 x 2	0 0 x 3	0 0 x 4	0 0 x 5	0 0 x 6	0 0 x 7	0 0 x 8	0 0 x 9
1	0 1 x 0	1 1 x 1	2 1 x 2	3 1 x 3	4 1 x 4	5 1 x 5	6 1 x 6	7 1 x 7	8 1 x 8	9 1 x 9
2	0 2 x 0	2 2 x 1	4 2 x 2	6 2 x 3	8 2 x 4	10 2 x 5	12 2 x 6	14 2 x 7	16 2 x 8	18 2 x 9
3	0 3 x 0	3 3 x 1	6 3 x 2	9 3 x 3	12 3 x 4	15 3 x 5	18 3 x 6	21 3 x 7	24 3 x 8	27 3 x 9
4	0 4 x 0	4 4 x 1	8 4 x 2	12 4 x 3	16 4 x 4	20 4 x 5	24 4 x 6	28 4 x 7	32 4 x 8	36 4 x 9
5	0 5 x 0	5 5 x 1	10 5 x 2	15 5 x 3	20 5 x 4	25 5 x 5	30 5 x 6	35 5 x 7	40 5 x 8	45 5 x 9
6	0 6 x 0	6 6 x 1	12 6 x 2	18 6 x 3	24 6 x 4	30 6 x 5	36 6 x 6	42 6 x 7	48 6 x 8	54 6 x 9
7	0 7 x 0	7 7 x 1	14 7 x 2	21 7 x 3	28 7 x 4	35 7 x 5	42 7 x 6	49 7 x 7	56 7 x 8	63 7 x 9
8	0 8 x 0	8 8 x 1	16 8 x 2	24 8 x 3	32 8 x 4	40 8 x 5	48 8 x 6	56 8 x 7	64 8 x 8	72 8 x 9
9	0 9 x 0	9 9 x 1	18 9 x 2	27 9 x 3	36 9 x 4	45 9 x 5	54 9 x 6	63 9 x 7	72 9 x 8	81 9 x 9

Name _____   Date _____

Enough for Everyone
Party Planner

$ _____

$ _____

$ _____

$ _____

$ _____

Item	How Many	Item Cost	Total
			Grand Total

Name _____   Date _____

Roll a Product Score Sheet

Directions: Take turns rolling two dice. Record rolls in factor boxes. Multiply factors to get the product. The player with the highest product in that round wins. Circle the winning product for each round. The player who wins the greatest number of rounds is the overall winner.

Player 1	Player 2

Player 1				Player 2		
	✕		=		✕		=	
	✕		=		✕		=	
	✕		=		✕		=	
	✕		=		✕		=	
	✕		=		✕		=	
	✕		=		✕		=	
	✕		=		✕		=	
	✕		=		✕		=	
	✕		=		✕		=	

Total Wins ☐   **Total Wins** ☐

The Great Big Idea Book: Math © 2009, Scholastic Teaching Resources

Product Cards

0	0	0	0	0	0	0	0	0	0
0	1	2	3	4	5	6	7	8	9
0	2	4	6	8	10	12	14	16	18
0	3	6	9	12	15	18	21	24	27
0	4	8	12	16	20	24	28	32	36
0	5	10	15	20	25	30	35	40	45
0	6	12	18	24	30	36	42	48	54
0	7	14	21	28	35	42	49	56	63
0	8	16	24	32	40	48	56	64	72
0	9	18	27	36	45	54	63	72	81

The Great Big Idea Book: Math © 2009, Scholastic Teaching Resources

Five in a Row

1. Stack the product cards together in a deck and shuffle.

2. Place the Five in a Row game board between two players.

3. Deal the product cards, giving each player $\frac{1}{2}$ of the deck. Keep cards facedown.

4. Have players take turns drawing a product card from their deck and placing it on a matching equation on the board.

5. The first player to completely cover a row of equations is the winner! This player does not have to have placed all cards in the completed row, but must place the final card that completes the row. The row can be horizontal, vertical, or diagonal.

Five in a Row

1. Stack the product cards together in a deck and shuffle.

2. Place the Five in a Row game board between two players.

3. Deal the product cards, giving each player $\frac{1}{2}$ of the deck. Keep cards facedown.

4. Have players take turns drawing a product card from their deck and placing it on a matching equation on the board.

5. The first player to completely cover a row of equations is the winner! This player does not have to have placed all cards in the completed row, but must place the final card that completes the row. The row can be horizontal, vertical, or diagonal.

Name _____ Date _____

The Great Multiplication Scramble

7	3	4	12	4	9	36	2	4	2	1	2
8	8	64	15	5	3	3	9	3	5	5	25
56	9	6	9	20	3	6	18	30	6	5	2
3	3	9	2	1	7	5	6	30	7	2	5
10	1	10	11	12	21	7	3	36	4	7	3
7	6	42	8	2	16	1	63	8	35	2	49
8	45	9	5	4	20	7	2	14	24	6	4
56	54	6	4	7	9	9	81	6	12	3	1
48	8	6	1	1	8	8	4	32	2	2	4
48	7	1	9	7	2	72	5	2	10	3	3
42	1	2	2	4	1	6	6	40	12	3	4
8	4	2	18	28	30	12	12	20	9	3	27

FACTS YOU FOUND

Example: 3 x 7 = 21

The Great Big Idea Book: Math © 2009, Scholastic Teaching Resources

Name _____   Date _____

Twice one are two,
Violets white and blue.

Twice two are four,
Sunflowers at the door.

Twice three are six,
Sweet peas on their sticks.

Twice four are eight,
Poppies at the gate.

Twice five are ten,
Pansies bloom again.

Twice six are twelve,
Pinks for those who delve.

Twice seven are fourteen,
Flowers of the runner bean.

Twice eight are sixteen,
Clinging ivy, evergreen.

Twice nine are eighteen,
Purples thistles to be seen.

Twice ten are twenty,
Hollyhocks in plenty.

Twice eleven are twenty-two,
Daisies wet from morning dew.

Twice twelve are twenty-four,
Roses, who could ask for more?

Name _____   Date _____

Fill in the Missing Product

x	0	1	2	3	4	5	6	7	8	9	10
0	0										
1								7			
2			4								
3					12						
4									32		
5						25					
6		6									
7							42				
8					32						
9									72		
10				30							100

The Great Big Idea Book: Math © 2009, Scholastic Teaching Resources

Measurement

When we want to find out how much there is of something, we usually count. How many pennies? How many children in the class? And so on. That works well with most things. But how about when you want to know how much flour you need for a recipe? Or how much wallpaper is needed to cover the wall? Or whether your luggage is too heavy for the plane? That's when counting no longer works and measurement comes in. With measurement, we find out how to quantify something, but in a very different way than by counting.

The activities are roughly divided into two sections: linear measurement; and weight, liquid, temperature, and mixed measurements. However, many activities overlap and combine more than one aspect of measurement, just as in real life. Many also have formats or ideas that you can easily adapt from one type of measurement to another. So, there is a great deal of flexibility.

One of the great things about teaching measurement is that it is so connected to real life. The NCTM *Principles and Standards* (2000) states, "Measurement is one of the most widely used applications of mathematics. Measurement activities can simultaneously teach important everyday skills, strengthen students' knowledge of other important topics in mathematics, and develop measurement concepts and processes that will be formalized and expanded in later years." Kids can see and experience measurement's immediate connection and application in their lives, from measuring flour for baking cookies, to weighing themselves, to finding out how long that inchworm really is. There are many natural connections to other areas of mathematics as well as to other curriculum areas.

Is It Longer Than . . . ?

A good way to begin an investigation into measurement is with simple comparisons. Remember when a clue in guessing games was, "Is it bigger than a bread box?" That probably wouldn't work too well today, but there are alternatives to the bread-box comparison.

- Begin by featuring an object that other objects will be compared to. For instance, you might hold up a board eraser and ask, "Is your pencil longer than this eraser?" Have students make some predictions, and then let them make an actual physical comparison.

- After a few examples, provide students with a copy of page 121. In the box near the top of the page, have them write the name of an object that they will use as their standard of comparison.

- Let students roam the room comparing objects with their standard, then use the two-column chart to record results (under Yes for things that are longer; No for things that are not longer).

- Have students share their results with the class. To reinforce learning, have students complete a second copy of page 121 as homework.

Nonstandard Desk Measurement

After simple comparisons, exploring nonstandard measures is a sensible next step. It helps students begin to experience and understand the need for standard measurement units such as inches or centimeters and how these standard measures help us communicate more easily about measurement.

Display a variety of manipulatives, such as multilink cubes, pattern blocks, and dominoes, as well as nontraditional manipulatives such as paper clips, twist ties, dried beans, and straws. If manipulatives are limited, have students work in pairs or threes.

Ask students to measure the width of their desks or tables using three different manipulatives. Model proper measuring technique, beginning at one edge and lining up the manipulatives end to end until you reach the other edge.

Have students record their findings on a copy of page 122. Discuss how, for example, while a desk might measure only 6 straws across, it may at the same time measure 60 beans. Which manipulative used the most units to measure across? The fewest?

This activity helps students gain an initial understanding of how we use a standard, such as a foot, for comparing and measuring. As students develop their knowledge in this area, add measurements to the exploration.

96

Measuring in "My Feet"

Help students appreciate the need for standard measurement by measuring the length of the classroom with their feet.

Have students spread out and take turns walking across the room heel-to-toe while they, or a partner, count the number of "feet" it takes to get from one side to the other. Have them record this number along with a tracing of their foot on a copy of page 123. Bring students together to compare measurements. An interesting discussion can result as to why some answers are the same and some are different. This is especially good if you, the teacher, do the same exercise and share your own answers. (You can add to the discussion by letting students measure and compare their foot tracings on their record sheets.) Now have students complete the same task using a ruler. Again, compare answers. Note how standardization results in more uniform answers and easier communication about what was measured.

Nanette Cooper McGuinness
Berkeley, California

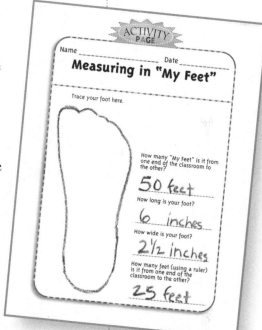

Literature LINK

How Big Is a Foot?

by Rolf Myller (Dell, 1962)

The King wants to give his wife a birthday present, but it's hard to find a gift for someone who has everything. He finally decides to give the Queen a bed and measures for it using his own large feet. The apprentice carpenter replicates the measurements he is given on a piece of paper using his own tiny feet. This is a recipe for trouble and a good lesson on the need for standard measurement.

Build Your Own Ruler

Children like to make things that they can actually use. Making their own measurement tools is naturally motivating and helps them understand the concept of how the tool works.

Introduce the concept of the inch and foot as standard measures. You may want to refer back to the nonstandard measurement activities and discuss how they help us see the need for standard measures. Give each student a copy of page 124. Point out the 12 separate inch segments on the paper. Have students cut out one and use it to measure a book, placing it on the book at one edge, then marking the spot with their finger, moving the inch segment up, and so on to complete the measurement. Have them record their answer, then use the tool to measure something longer, such as their desk. Finally, give them each a piece of oaktag a foot long. Have them glue on the 12 separate segments end to end to make a foot-long ruler. Show students how to mark off and number the inches, then let them use the tool to make new measurements. Is it a little easier than moving that little one-inch piece around? Letting children discover the usefulness of tools in this way really helps them appreciate and understand the measurement tool.

ENGLISH Language LEARNERS

This is a "handy" way to help English language learners, and other students, get a "handle" on measurement terms. Make copies of the template on page 125. As you study various units of measure, give students a copy. Guide students in completing the information, including what attribute the unit measures, measuring tools that represent that unit, and so on. Have students draw a picture of something they could measure with the unit and the measuring tool they would use. Punch holes in these pages and tie with yarn. Add pages as you introduce new units of measurement. Students can easily refer back to units of measure they've completed pages on, but the real purpose is reinforcing the idea as they complete their handy pages.

Snaky Measurement

This activity lets children try out their new rulers and become more accurate in the measurements they make.

- Cut out five or six paper "snakes" from roll paper. Keep the snakes straight (don't cut curves), and cut them in different lengths, based on measurements your students are working with.

- Number the snakes, then place them on the floor around the room. Have children get paper, pencil, and their rulers (see Build Your Own Ruler, page 98), then arrange themselves in groups of three or four next to a snake.

- Borrow a ruler from a student and demonstrate how to place it end to end to measure a snake. Ask students in each group to estimate, then measure, the length of their snake, and record both on their paper.

- When students in each group have estimated and measured their snake, have them rotate to a new snake and repeat the activity. Continue until each group has estimated and measured each snake. Then bring the class together and compare results. Discuss reasons for any differences in the measurements of each snake. (For example, when students are placing the ruler end to end to measure a snake, the ruler might not be positioned exactly the same each time.) Notice whether students' measurements become more accurate with practice.

Bonnie Bliss-Camara and Judi Shilling
Dutch Neck School
Princeton Junction, New Jersey

Literature LINK

How Tall, How Short, How Far Away

by David Adler (Holiday House, 1999)

This book is a good introduction to linear measurement. With bright, engaging illustrations it explores in a very clear, easy-to-follow manner the measurement systems of ancient Rome and Egypt. It shows how these made clear the need for standardization, and then makes connections to modern measuring situations as it looks at the English and metric systems.

How Big Are Those Feet?

Kids love dinosaurs, partly because dinosaurs were SO big! Capitalize on this with a dinosaur measurement investigation.

⑤ To prepare, get a large sheet of cardboard or roll paper and create a model Seismosaurus footprint. The Seismosaurus footprint should measure about 36 inches wide and 36 inches long. It will have five toes with the two on one end being about five inches longer than the other three.

⑤ After students have gone for the day, put the footprint in the middle of the floor so it is the first thing they see when they return the next morning. Upon their arrival ask, "Can anyone guess what this might be? How could we measure it?" Divide the class into groups of two to four students and let them take turns over the next few days exploring the print through measurement. Give children copies of page 126 to record their measurements, and invite them to share their results with the class.

Carol Kirkham Martin
Piney Ridge Elementary
Sykesville, Maryland

TIP

A good comparison to recommend is to create a paper template of a child's footprint and use that as a nonstandard unit of measure to see how many would fit in the Seismosaurus print.

Literature LINK

Is a Blue Whale the Biggest Thing There Is?

by Robert E. Wells (Whitman, 1993)

After children have measured the Seismosaurus footprint every way they possibly can, share this book to explore more big things. The book's wonderful visual comparisons make it easy for children to picture just how big such things as whales, Mount Everest, and the sun are. For example, it shows how many whales it would take to be as "big" as Mount Everest and then how many Mount Everests it would take to be as "big" as the sun. Follow up by letting children use a piece of string to compare the length of a whale with other things. For example, how does it compare to your school? to a car? to your whole class laid head to toe?

Cotton Ball Olympics

When introducing smaller units of measure—such as centimeters, inches, or half-inches—challenge students to participate in the Cotton Ball Olympics and measure their results.

Provide each student with a cotton ball and divide the class into groups of three or four. The Olympics can be held inside or outside. Run simple events like the Cotton Ball Throw. Students stand behind a line and throw the cotton ball overhand as far as they can. The student then measures the distance from the line to where the ball has landed and records it. Everyone takes a turn and then tries the next event. Other events could include the Underhand Cotton Ball Throw, Cotton Ball Kick, Cotton Ball Blow, and any other creative events you can come up with. The point is, the cotton ball isn't going too far, so the measurements that are created are going to be small and very appropriate for smaller units such as centimeters and inches. When everyone is finished, you might want to tally all the results in chart form to see the range of results and maybe hand out those Olympic medals!

Choose the Best Unit

Can measurement be linked to grammar? You bet!
Try this activity to reinforce learning in both areas.

Discuss with students the definition of a noun. (*A noun names a person, place, or thing.*) Brainstorm a list of nouns as a class and record these on the board. Now ask students to help you list units of measurement that would be appropriate for each noun. In other words, if you were going to measure attributes of that noun, what would be reasonable units to measure in? For example, if someone mentioned *car*, reasonable units of measure might be pounds, feet, yards, or meters. For *paper clip*, better units might be centimeters, ounces, or grams. List the nouns and their suggested measures on the board. Take it a step further and have students estimate what the actual measurements might be for the nouns. For homework, ask students to pick one noun, then measure the object and share the results with the class. Compare actual measurements with estimates.

The Biggest Bug Ever

Does measurement bug you? That's a good question to ask your class as you begin this project.

Give each student a large sheet of drawing paper, at least 11 by 17 inches. Tell students they are going to create the biggest bug ever. It can be a bug of their own invention or a real insect they know of, such as an ant or fly, but "the biggest" ones imaginable. Provide paint, markers, crayons, colored pencils, paper scraps, and whatever other art materials are available and let students create their bugs. Students can name their bugs if they wish, then record information about them, beginning with some measurements. Have students measure and record the length and width of their bugs. Have them then estimate, based on what they know about things of that size, how much their bug might weigh if it were real.

TIP

As an extension, have students bring their papers home. Encourage them to share the poem with their families, then find more things at home that measure an inch and record them on the backs of their papers. Compare findings in class.

The Inchworm's Trip

Ask students if they have ever seen an inchworm. Let them have fun imitating the interesting way it moves, then ask: "Do you think inchworms are really an inch long?" Try this activity to explore the question and learn more.

- Give each student a copy of page 127 and read the poem together.

- If it's the right season and you have inchworms in your area, invite students to catch one, give it a safe, temporary home in a bug jar, and bring it in to share. Measure it. Is it an inch long? If you don't have access to inchworms, share a picture and research the answer to the question.

- After exploring inchworms (or pictures of them), ask students to find items in the classroom that are about an inch long. Have them record their findings on their papers as they go, then share results with the class.

Crooked, Curvy, Straight

After you have done some work with linear measurement, try this fun and challenging exploration with students.

Wait until students have left the room, then use tape to make a straight line on the floor. (Use a length you would like students to work with.) Have students work in groups to measure the length of the line and record it. The next day, surprise them with a crooked lighting-bolt style line next to the straight line. The segments of this line should add up to the same length as the first line. Discuss ways students can measure the new line, then let them measure and record the length. On the third day, tape down a curvy line, the same length as the first two. Have students brainstorm how they can measure this line. (Two common solutions are using centimeter cubes end to end or using a length of string, then measuring the string.) Again, have them work in their groups to measure and record the length. Let groups discuss their methods, and compare their results.

Test-Taking Tip

Regular, guided practice with rulers helps students develop strategies for solving measurement problems on tests.

Many standardized tests include a section on measurement that has students use a punch-out ruler or ruler provided by the teacher. This ruler is usually marked in centimeters on one edge and inches on the facing edge. To help students use the ruler correctly, provide practice measuring in both centimeters and inches. Have students note the difference between them, particularly as to how they are not equal or even roughly equal. Provide plenty of practice measuring from the end of the ruler, too. Sometimes students will line up the object on the test page with the 1-inch mark on the ruler. (You start counting with "1," right? So, many students assume you start measuring from "1" as well.) This is a common mistake, and there will most likely be a distractor answer that features the measurement students would arrive at if they made this error.

For an interesting extension, ask students to place the balls in order by how much they think they weigh. Let them use a scale to compare actual weight to their estimates. Discuss how some objects, though smaller, can weigh more than bigger objects.

Have a Ball Measuring

Children use balls constantly, so they make a very natural connection to learning about circumference. You can explain circumference simply as the distance around a circle or sphere.

- Bring a variety of balls into class—for example, a golf ball, whiffle ball, softball, beach ball, soccer ball, and basketball—and invite students to do the same.

- Have children arrange the balls in order from the one they think has the smallest circumference to the one with the largest. Record the order on the board. Ask children to make estimates of the circumference of each ball and record those estimates on paper.

- Let children work in groups to measure the circumference of each ball using a tape measure. Have them record the measurement on the board under their ball. Is the original order correct? How do students' estimates compare with their measurements?

Janet Worthington-Samo
Saint Clement School
Johnstown, Pennsylvania

Meter Monsters

The concept of a meter is a new idea to many students. Familiarize them with this unit of measurement by making Meter Monsters.

Cut adding-machine tape into meter-size strips and give one to each child. Invite children to draw a monster on their strip, making sure the monster stretches from one end to the other. (This is going to be a long, thin monster!) If possible, laminate the monsters. Now see if students can find objects in the room, halls, or playground that are about a meter long. How about two meters? three? How many meter-monsters long are they?

Jacqueline Clarke
Cicero Elementary
Cicero, New York

Measurements About Me

It's interesting to find out about ourselves and measurement is one way to do that. This particular activity is a good way to get students involved with metric measurement.

◎ Form groups of four or five students and make a copy of page 128 for each student. Give each group a meterstick and length of string about six feet long.

◎ Have students work together in their groups to measure each other and complete the record sheet. First, they should estimate a body measurement. For example, the first measure is from belly button to feet. They should measure this distance with string, marking the string with a finger or colored marker, then measure the length with the meterstick. Have them record each length on the record sheet in metric units.

◎ The next day, give each student a sheet of roll paper large enough to trace their body on. Have partners trace each other's outlines on the paper. Using the measurement information and metersticks, students can add details to their drawings in the proper places—for example, they'll know where the chin and belly button go by measuring with the data they have.

◎ Invite students to use their artistic abilities to add features, clothes, color, and other extras. These finished "bodies" make great hallway displays. For Back-to-School Night it's always fun to see if parents can pick out their "child" from among the measured paper students.

Wendy Wise-Borg
Rider University
Lawrenceville, New Jersey

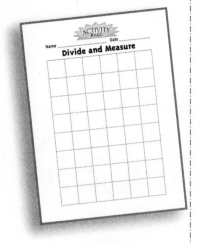

Divide and Measure

Here's an effective way to begin exploring perimeter and area.

- ⟲ Give a copy of page 129 (a grid of 1-inch squares) to each student.

- ⟲ Have students follow along on their grids as you introduce the concepts of perimeter and area and show how to measure them on the paper.

- ⟲ Direct students to cut the grid in half and find the new perimeter and area. Have them cut it in half again and find the new measures. Now have students experiment with the pieces, recombining them to create new shapes. Have them measure the perimeter and area each time.

- ⟲ Ask: "Can two shapes with the same area have different perimeters?" Have students prove or disprove their statement with their models. When students find a combination of shapes they like, ask them to glue the pieces down on a sheet of paper and record the area and perimeter.

Janice Reutter
Medicine Lodge Primary School
Medicine Lodge, Kansas

Paper Strip Sculptures

This art project demonstrates once again that even though you may work with the same materials and limits, the outcomes are endlessly varied and creative.

Precut some 1-inch-wide colored construction paper strips. Make each strip a different length—from 2 to 12 inches. Explain and demonstrate how these flat paper strips can be used to create three-dimensional sculptures. For example, bend one strip at the bottom to form a foot. Do the same on the other end. A little glue on the bottom of each foot allows the strip to be easily stuck to a base. (Use a sheet of oaktag.)

TIP

For an exciting extension, invite students to combine their pieces for even larger and more challenging models and resulting perimeters and areas.

Give each student a 9- by 12-inch sheet of colored construction paper to create a paper strip sculpture. Explain the measurement requirement: They can use 12 strips to create their sculpture. Each strip must be a different length, from 2 to 12 inches. Give students a base to glue their sculpture to—oaktag or cardboard is recommended. Strips can be decorated or shaped in various ways by cutting, but they must be carefully measured and cut to the required lengths before altering or attaching. It's interesting to see the results and the creativity students put into this project.

Playground Architects

For most children, the playground is one of the best places at school. Here's an opportunity for kids to explore area and perimeter in a creative way while enjoying this favorite place.

- Take a playground break with your students. As they play, encourage them to notice positioning of playground equipment, what's there, what they like, what's missing, and so on.

- Back in class, give each student a copy of page 130, a grid of 1-centimeter squares. Demonstrate how on a piece of this paper, you can sketch in play areas such as a basketball court. Show how you can draw each part of the playground by showing an aerial or bird's-eye view. Also explain that what you draw on the paper will be squared off at the edges, not rounded.

- Have students sketch out their own plans for a playground, then count the squares inside each section of the drawing to find the area of each playground compo-nent, and add the sides to determine perimeter.

Wendy Wise-Borg
Rider University
Lawrenceville,
New Jersey

TIP

For an added challenge, have the class determine the total area used by their playground components. With that information they can also figure the amount of unused area.

Does It Weigh More Than . . . ?

This activity makes a good independent learning station or introductory lesson to learning about weight.

Set up a weigh station in your classroom. Provide a balance scale or bathroom scale, a group of four or five different objects, a target object for students to compare the other objects to, and record sheets. To make the record sheets, fill in the target object at the top of the reproducible on page 131 and photocopy a class set. Have students take turns at the station, using the scale to compare weights of the target object and other objects. For each object have students answer the question, "Is it heavier than the (target object)?" For an additional challenge, ask students to order the items from heaviest to lightest and show where the target item falls in the list. (They can do this on the back of their record sheets.) Change the objects at the weigh station weekly to provide additional practice and keep the activity fresh. As you move the class into standard measures, the target object can become a one-pound weight or similar standardized weight.

Nonstandard Weight Exploration

Marbles add a playful element to this exploration of nonstandard measurement in weight. Use this activity as an introduction to using balance weights like grams, ounces, and pounds, and helping children see that the standard measures make sense.

⊙ Set up a balance scale and place a bowl or container on each side. Explain that students will be measuring the weight of objects from around the room—in marbles! To demonstrate, place an object, such as a book, in one pan. Then begin to place marbles in the other pan. Add one marble at a time until the pans balance. Count the marbles to see how many marbles the item weighs.

⊙ Repeat the activity with a different object. This time, ask students how many marbles they think the object will weigh. Let them feel the weight of one marble first, and then a handful of marbles, so they have some background knowledge to draw on.

⊙ After trying this a few times, show students a larger marble. Ask them if using a pan of these marbles instead of the regular marbles would give them the same result—obviously it would not. Students might recognize that this is the same problem that is seen in linear measurement when measuring with human feet. One person's foot is not the same as another person's. (See page 97.) This is true for marbles or any other nonstandard measure.

Now let children try the activity on their own, working with partners to measure objects around the classroom in marbles. Have them record their results. When everyone has had a chance to make some measurements, bring students together to share and compare results.

Make a Scale

Here's another opportunity for students to construct and use their own measuring tool.

Divide the class into groups of three or four. Provide each group with the materials. (See list at right.) Guide students in following these steps to make a scale. (See illustration for further information.)

1 Place the ruler on a desk, with about half of it hanging over edge of the desk. Anchor the ruler to the desk by setting the heavy book on it.

2 Loop the rubber band over the end of the ruler that hangs over the desk.

3 Thread a piece of string through each hole of the box and tie a knot. Knot the four strings together in the center, then tie the ends of the strings to the end of the rubber band.

Have students tape a piece of drawing paper to the side of the desk alongside the scale and mark where the bottom of the box is. As students place an object in the box, the box will lower. If they mark where the bottom of the box is on the paper now, the difference between the two marks will indicate the weight.

Help students try a variety of objects and mark where the bottom of the box falls each time. This will provide good weight comparisons in a simple way. Compare the differences in marks.

Help students understand that the lower the mark, the heavier the object. Remind students to always estimate first and then check their predictions.

Materials

- a ruler
- a heavy book (such as a dictionary)
- a long rubber band
- four pieces of string (each about a foot long)
- a small cardboard box with a hole punched in the center of each side near the top

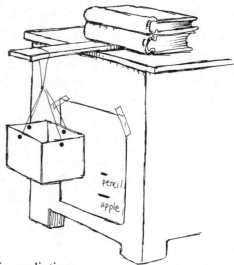

109

When your "eye-popping" demonstration is complete, you can use the kernels and balance scale to weigh other objects in terms of popcorn kernels. For example, a pencil might weigh 20 kernels. An eraser might weigh 25 kernels. This is an interesting way to explore non-standard measurement in the area of weight.

Popcorn Comparison

According to recent data, popcorn is the number one snack food in the United States. It is also useful here in learning about weight.

Take two handfuls of unpopped popcorn. Count the kernels to make sure you have an exact match. Weigh them on a balance scale in front of the class to demonstrate that they weigh approximately the same. Now pop one handful in a hot-air popper, or substitute the same number of already popped kernels. Weigh the popped kernels against the unpopped on the balance scale. Ask students which weighs more now and what they think happened. (*The interesting thing is that when popcorn pops, moisture is released. Once popped, you are no longer weighing that moisture, so the popped corn weighs less.*)

Sue Lorey
Grove Avenue School
Barrington, Illinois

Literature LINK

Measuring Penny

by Loreen Leedy (Henry Holt, 1997)

In this wonderfully illustrated picture book, Lisa gets a very open-ended homework assignment from her teacher, Mr. Jayson. He asks the class to go home and measure something. Anything. And to measure it in as many ways as they can: height, width, length, weight, volume, temperature, and so on. Lisa goes home and decides to measure Penny, her dog. And she does it in every clever way she can think of, using both standard and nonstandard measures. After reading this story, give your students the same homework assignment. They'll have fun sharing their results at school.

The Heaviest Book in School

Students will often talk about how heavy their books are, but which of their books is actually the heaviest? In fact, which book is the heaviest in the entire school? This is a fun and motivating investigation that provides students with practice in weighing, comparing, and graphing.

- Begin by having students work in teams of three to measure books they have in their desks. Have groups share their findings with the class. Record the heaviest book.

- Now expand the search to books in the room. Record the heaviest.

- The next day, have students scour the school for more books. Have each child bring books from anywhere in the school. (Forewarn the librarian!)

- Compare all results and weigh the finalists to determine the heaviest book in the school.

- For more fun, try the challenge in reverse and attempt to find the lightest book in the school.

Mystery Measurements

A cloaked mystery object will pique students' interest, as they try to determine what's underneath, based on measurements you provide.

Before the class arrives for the day, take an object in the room, such as a book or pencil sharpener, and measure as many attributes as you can. For instance, you could measure the length, weight, and width of the pencil sharpener. Now, place the object on a desk and cover it with a mysterious opaque cloth. (Black is good.) When the class returns, explain that there is a mystery object under the cloth and that you would like them to guess what it is. Give them the first clue—a measurement you have recorded—and invite guesses. If no one guesses it, give the second measurement clue. Continue giving measurement clues until students figure out what the object is. Repeat the activity often, using a variety of objects that include a good contrast of heavier and lighter, longer and shorter objects.

Interactive Morning Message:
Measuring Tools Brainstorming

To start the day, and put measurement on everyone's mind, try this morning message activity.

Place a measurement tool such as a scale, ruler, yardstick, meterstick, measuring tape, or measuring cup next to your morning message pad, and in your message to students ask what they could measure with the tool. Invite students to record their ideas on the morning message paper. Take time during your class meeting or some other time to discuss the different ideas and to help students build their understanding of what the possibilities are with a given measurement tool.

Ready for the Trip

Here's a simple simulation that provides a real-life connection for weight measurement.

- Tell students you will be taking a simulated trip on an airplane to an exotic place (perhaps somewhere you are studying about in social studies). Send home a letter telling students to pack a back-pack with items they need for an overnight trip. Have students bring their packed backpacks to school. (It's a good idea to note to parents in the letter that students really need to do this part.)

- When students bring in their backpacks, have them use bathroom scales to weigh them. Decide which units you want the class to report in (pounds and ounces or kilograms and grams).

- Have students add together all of the weights in their group. Record this data on the board and ask each group to figure out how much baggage weight the total class has for the plane. Whatever the result is, inform the class that unfortunately they are overweight on the baggage. Give them a weight limit that is 50 to 100 pounds less.

- Now have students in each group make an adjustment by removing items from their backpacks and weighing again. Do they make the weight limit this time? If not, have them remove more items.

- Once the total is under the weight limit, you can "fly" your class to their destination. Set up chairs in rows of six with an aisle space down the middle to resemble airplane seating. Have students board and stow their bags under their seats. Simulate take-off and show a video about their destination at the front of the "plane."

Interactive Morning Message:
The Estimation Object

An object on a table or desk next to your morning message invites estimating and measuring.

In your message, ask students to estimate a particular measurement attribute of the object—for example, length, width, or weight. Tell students they can look at the object, but use only their observation skills to give their best estimate. Have students record their estimate on the morning message sheet. Bring students together and look at the range of estimates. Circle the highest. Underline the lowest. Now, get a couple of helpers and measure the object for the given attribute. Measure again to make sure. Compare the estimates and the actual measurement. How close did students get? How close will they get next time? The answer is, even closer with practice!

Class Pet Measurement Journal

If you have a class pet, try this ongoing activity to enhance student learning about the pet and about measurement.

Make a measurement journal master with space for recording the following data: Date, Pet Keepers, Height/Length, Width, Weight, Other. Make multiple copies of the journal page and staple or place it in a binder to make the journal. Schedule students in pairs to measure the class pet and record the data in the journal. They should note their names, the date, the pet's weight, height, width, and any other relevant measurements. Chart data in graph or table form and review it regularly to see if the pet is growing.

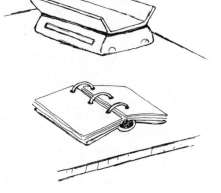

Which Holds More?

A good way to begin exploring the idea of liquid measurement is to raise students' awareness of liquid measurement around them.

Send home a note with students asking them to bring in clean, empty, plastic liquid containers. (Have students leave any labels intact.) After you have collected a good assortment of containers, have students try to arrange them in order from the greatest capacity to the smallest. Record the information from the labels on a chart and discuss measurements mentioned on the labels. This is a good pre-assessment activity that helps you begin to find out what students know about liquid measurement.

The Coffee Shop Problem

Here's a little demonstration to help students see the need for standard measures in liquid measurement.

- Prepare by bringing in four different-size cups. The best choices are a large, soup bowl–like cup, a teacup, an ordinary coffee mug, and a cup from a child's tea set. Put these in a bag. On chart paper or the board write, "Coffee Shop. Coffee 5¢ a cup."

- Gather students together and explain that you are opening up your pretend coffee shop. Select four students to act as your customers. Instruct them to come in the shop and, one at a time, ask for a cup of coffee. Pretend to take a nickel from the first customer. Remove the coffee mug from the bag and pretend to pour coffee into it. Give the coffee to the customer and have that student step aside.

- Continue taking pretend nickels and pulling cups from the bag and pouring. Once all four students are standing at the front with their different cups, ask the class if everything is fair here at the coffee shop. The discussion that ensues will focus on the need for standard measurement—in this case, standard liquid measurement—so that when we say "cup," we all know what we are talking about.

Tests sometimes ask children to measure objects with a length or weight that is not exact—for example, "About how long is the leaf?" *About* is the key word. Let students practice with measuring objects that do not fall exactly on a line on the ruler or on a scale—not only because they may be tested on such a skill, but because very few actual objects measure to exact whole numbers on common measuring tools.

Fill the Gallon

Simple card games are fun, can be played anywhere, and help reinforce math ideas in an active, engaging way. "Fill the Gallon" is a good example.

Make a copy of page 132 for each student. Have students glue this page to oaktag, then cut apart the cards. To play, two players place a deck of cards facedown between them. Players take turns drawing a card from the deck (one card per turn), trying for a combination of cards that will equal one gallon. For example, a player who has a quart card, a half-gallon card, and another quart card has "filled a gallon." The first player to "fill a gallon" wins.

Taking the School's Temperature

Do your students ever make claims about the temperature at school, saying such things as, "The gym is always sooooo cold," or "Our classroom is the hottest room in this school"? Try this activity to see what the truth of the matter is.

- Explain to students that they will be investigating claims about the temperature in various parts of the school. Divide the class into groups and give each a small thermometer. Assign each group a room somewhere in the school. Arrange a time for them to go to the room and talk to the teacher there about the investigation, asking for permission to tape a small thermometer near the door.

- Have students check their thermometers at an assigned time each day to record the temperature. It's important that everyone go on the same days and the same times.

- At the end of two weeks, compare temperatures from the different locations. It may be useful to graph this information or create a table for easy comparisons. Now the class can authoritatively answer where the hottest and coldest places at school are.

Another game that can be played with the same cards on page 132 is Liquid Measurement War. This is played just as regular War is played. In each round, the card with the largest liquid measurement is the winner of that round.

The Crazy Concoction

Recipe writing is a natural way to combine writing and measurement. Recipes combine numbers, measurement, and sentences describing actions (*mix the batter, put it in the oven*). Use this activity to whip up some mouth-watering math in your classroom.

- Share a favorite recipe with students (you might place it on an overhead projector) and guide them in noticing the various elements such as time, temperature, tools, ingredients, liquid measures, and weights. Prepare the recipe if possible, so that students have hands-on experience with following a recipe.

- To give students a little more experience with following recipes, invite them to ask a family member to help them prepare a recipe at home. (This is an optional assignment.)

- Now, for some writing fun, invite students to create a "Crazy Concoction" recipe. They should write their own recipe, which can include just about anything (nothing too rude or disgusting), but must have all the conventional elements of the regular recipe, including quantities, measurements, time, and so on. Have students give their recipes names and draw a picture of the finished product. Their recipes will make for a "tasteful" display!

Measurement Assessment Station

When assessing a student's understanding of measurement, try setting up a one-on-one measurement assessment station.

To assess student understanding of liquid measures, use a container of water as well as several measuring cups. Ask a student to measure different amounts of water and record his or her answers for you. This will take only a few minutes and you will be able to observe the accuracy of the student's measures and, more importantly, the methods he or she uses to measure. The same procedure works well for assessing understanding of linear measurement.

Clay Work

Measurement lends itself very well to kinesthetic learners—it is very hands-on. To make it even more hands-on, introduce clay into the lesson.

Give each student some clay. Have students use the clay to make an object of their choice. When they're finished sculpting, have students record measurements about the piece, including weight, length, and width. You can work this from the other direction, also, by giving out the clay and measurement parameters first. For example, you might say that the object must be between 4 and 5 inches long, must weight less than 8 ounces, and so on. In this case, students should then measure and record along the way and at the end to prove that their objects actually meet these criteria. It is interesting to note that the objects can be very different from each other, but still be in the same measurement ranges. This makes a good learning center or independent activity.

All Sports of Measurement Bulletin Board

Children at this age are very interested in sports. As a class, compile a comprehensive list of sports. Then try this activity to explore real-life measurement.

Ask students to work independently or in small groups to record how measurements are used in the sports they named. Examples might include yards for a first down (football) or feet between bases (baseball). Bring students together to share findings. Record them on the board or chart paper, then have each student choose one of the sports and measurements and explain the measurement connection. Students can add illustrations, then display their work to create a fun and informative bulletin board.

Sue Lorey
Grove Avenue School
Barrington, Illinois

Weights and Measures Graphic Organizer

Sometimes remembering all of the weights and measures in the English and metric systems can be a challenge. The handy graphic organizer on page 133 can help.

⟲ Give each child a copy of the organizer. It includes all the standard measures students would most likely use at this grade level, as well as abbreviations.

⟲ Have students tape it in a notebook, on a desk, or in some other convenient place for ready reference.

⟲ Familiarize students with the information on the organizer with a quick game. For example, ask: "What does the abbreviation *ml* stand for? How many ounces in two pounds?"

⟲ Let the student who answers the question ask the next one. As your study becomes more advanced, add information about additional weights and measures to the organizer.

ENGLISH
Language
LEARNERS

For many students coming into the United States from other countries, there are new cultural systems to learn about in addition to a new language. This includes our measurement system. The system of inches, feet, yards, pounds, quarts, and so on is unique to our country. Your English language students are most likely used to the metric system. To help students adjust to this new system of measurement, you may want to send home a copy of page 133, the Weights and Measures Graphic Organizer. Attach a note to families, mentioning that the class will be learning about both systems, and you want them to have this at home as ready reference.

Take-Home Activity:
Measurement Makes News!

Connecting student learning to home and to real application pays great benefits in building student understanding and fostering the home-school connection.

Give each child a copy of page 134. Explain to children that they will be completing the page for homework. Review the assignment, which asks children to look for examples of measurement in a newspaper or magazine and to complete some information about it. Encourage children to invite a family member to work with them. Have them return the papers to school by a specified date, then use the information they gathered as discussion starters: "Did anyone find examples of measurement in sports (for example, how far a football player ran to make a touchdown, or how fast a runner ran a race)? How about examples of measurement in other news, such as how many inches of rain fell in a storm? In how many different areas of the news did students find examples of measurement?" When you wrap up your discussion, use the papers to create an informative and fun bulletin board. Make additional copies available for students who want to find more measurements in the news.

Take-Home Activity:
Recipe Math

More math probably takes place in the kitchen than anywhere else in the house. Tap into this with a fun take-home activity.

Give children copies of pages 135 and 136, which include a recipe children and families can make together, and a comment sheet so they can reflect on the math involved and share a favorite recipe of their own. (See Tip.) Remind children not to use the stove or oven on their own. As students prepare the recipes with their families, invite them to share their experiences, particularly with measuring. What ingredients did they measure? What tools did they use?

Photocopy recipes your students' families share. Put them together to make recipe books children can take home. They'll enjoy trying their classmates' favorite dishes—and will gain more measurement practice in the kitchen each time!

Take-Home Activity:
Measurement Kit

Here's a fun kit to send home with students. They'll look forward to its being their turn to take home the kit—and to exploring measurement at home with their families.

Place the following items in a reclosable bag:

- ruler
- tape measure
- laminated copy of page 137
- fresh copy of page 138 (replenish each time you send the kit home with a child)

Review the contents of the kit with students. Let them name each measuring tool they see and tell how to use it. Read over the list of items students will measure at home (door, fork, table, window, refrigerator, bed), then let them take turns naming other things they could measure at home with the ruler and tape measure. (You can easily make changes to the six items children will measure. Simply white out on a master copy the items you wish to change and write in new items.)

Send home the kit with the first child, and when it is returned, discuss the measurement data gathered. Record the data for the first six measurements (the ones everyone makes) on a class comparison chart. Each time you add new data to the chart, use it to make comparisons: "Is everybody's door the same width? Does it appear that there are some standard sizes out there in the real world? What does the data tell us?" As the kit makes its rounds through the class, the chart data will grow, and students will have more information on which to base their answers.

Put together a few kits so that there's less waiting time for each child to take one home.

Name _____ Date _____

Is It Longer Than . . . ?

YES	NO

Name _____

Date _____

Nonstandard Desk Measurement

Object Measured	Unit of Measurement	How Many?
Example: desk (length)	lima beans	36

Name _____   Date _____

Measuring in "My Feet"

Trace your foot here.

How many "My Feet" is it from one end of the classroom to the other?

_____

How long is your foot?

_____

How wide is your foot?

_____

How many feet (using a ruler) is it from one end of the classroom to the other?

_____

Name _____   Date _____

Build Your Own Ruler

1"
1"
1"
1"

Name _____   Date _____

Measurement Handbook

Unit

Attribute

Measuring Tools

English or Metric?

Your Name

Name _____   Date _____

How Big Are Those Feet?

Show four ways that you measured the dinosaur footprint.
Record your measurements.

1. My measurement:

_____

2. My measurement:

_____

3. My measurement:

_____

4. My measurement:

_____

KEEP GOING!

Can you think of more ways to measure the Seismosaurus footprint?
Show them on the back of this paper.

The Great Big Idea Book: Math © 2009, Scholastic Teaching Resources

Name _____   Date _____

The Inchworm's Trip

Inch,
by inch,
by inch
he crawls

through our classrooms,
through our halls.
He's one inch long
and will not stop,
Inching, inching,
past the mop.

Inching, inching,
up my chair.
Now he's inching
through my hair—

His way of "walking's"
fun to see—
Does it tickle him
as much as me?

—by Sandra Liatsos

"The Inchworm's Trip" from POEMS TO COUNT ON: 32 TERRIFIC POEMS AND ACTIVITIES TO HELP TEACH MATH CONCEPTS by Sandra Liatsos. Copyright © 1995 by Sandra Liatsos. Reprinted by permission of Scholastic Inc.

TRY THIS!

Color and cut out this inchworm ruler. Use it to measure around your home. What measures an inch? What is shorter than an inch? Longer than an inch?

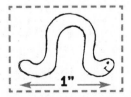

1"

Name _____ Date _____

Measurements About Me

	Estimate	Measurement
belly button to feet		
chin to belly button		
knee to ankle		
top of head to chin		
top of head to bottom of feet		

Name _____   Date _____

Divide and Measure

Name _____ Date _____

Playground Architects

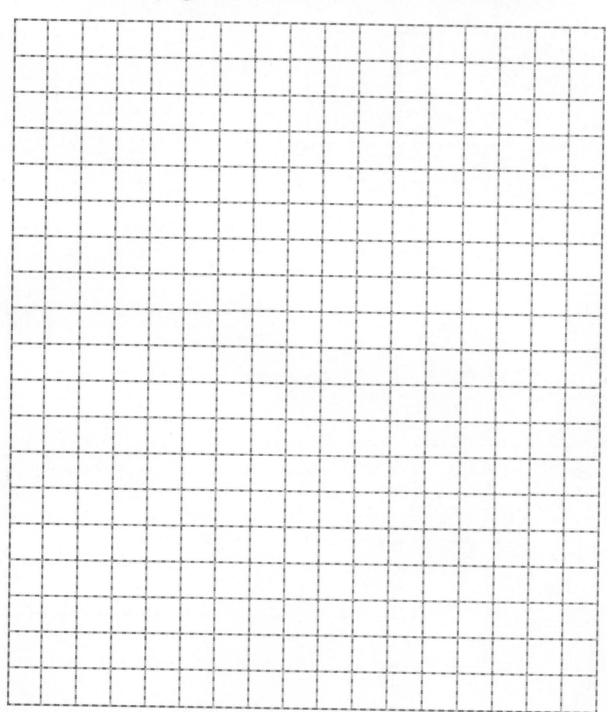

The Great Big Idea Book: Math © 2009, Scholastic Teaching Resources

Name _____   Date _____

Does It Weigh More Than . . . ?

YES	NO

Fill the Gallon

cup	cup	cup	cup	cup	cup
pint	pint	pint	pint	pint	pint
cup	cup	cup	cup	cup	cup
quart	quart	quart	quart	half gallon	half gallon

Name _____ Date _____

Weights and Measures Graphic Organizer

English	Metric

Length
inch (in)
12 inches = 1 foot (ft)
3 feet = 1 yard (yd)

Length
centimeter (cm)
100 centimeters = 1 meter (m)

Weight
ounce (oz)
16 ounces = 1 pound (lb)

Weight
gram (g)
1,000 grams = 1 kilogram (kg)

Liquid
fluid ounce (fl oz)
8 fluid ounces = 1 cup (c)
2 cups = 1 pint (pt)
2 pints = 1 quart (qt)
4 quarts = 1 gallon (gal)

Liquid
milliliter (ml)
1,000 milliliters = 1 liter (l)

The Great Big Idea Book: Math © 2009, Scholastic Teaching Resources

Name _____   Date _____

Measurement Makes News!

Dear Family,

We are learning about measurement. Measurement is a very important part of mathematics and life. You can strengthen your child's measurement skills by completing this assignment.

- Help your child look through newspapers and magazines to find examples of measurement. For example, there might be a news story from a sports page that mentions how far a football player ran for a touchdown. It may even mention the player's height and weight. Any examples that include measurements are good, but the best are those that feature things your child is interested in.

- Cut out the picture or article. Work with your child to circle the measurements.

- Complete the remaining information and discuss why this information is helpful and/or interesting. Please have your child return the assignment by _____ .

Sincerely,

Measurements (pounds, inches, liters, etc.) _____
_____

How the measurement was used: _____

_____

The Great Big Idea Book: Math © 2009, Scholastic Teaching Resources

Name _____ Date _____

Recipe Math

Dear Family,

Your child is learning about measurement. Measurement is an important part of learning mathematics, and it is also a big part of our lives—particularly in the kitchen! Here's a recipe that you can make with your child. Allow your child to do as much of the measuring as you are both comfortable with. Please complete the attached sheet with your child and return it to school.

Sincerely,

Oatmeal Cookies

1 cup butter

1 cup brown sugar

$\frac{1}{2}$ cup white sugar

2 eggs

1 teaspoon vanilla

1 $\frac{1}{2}$ cups flour

1 teaspoon baking soda

1 teaspoon cinnamon

3 cups oats

1 cup raisins

$\frac{1}{2}$ cup grated carrots

Heat the oven to 350° F. Beat together butter and sugars. Add eggs and vanilla. Beat well. Add flour, baking soda, and cinnamon. Mix well. Add oats, raisins, and carrots. Mix well. Drop by teaspoons on an ungreased baking sheet. Bake for about 10 minutes until golden brown. Makes about 3 dozen.

Name _____ Date _____

Recipe Math
Response Sheet

1. Who helped make the cookies? _____

2. What measurement units did you use to make the cookies? (cups, teaspoons, etc.) _____

3. What was your favorite part of the measuring?
   _____

4. What did you learn about measuring? _____

5. Fill in the face to show how you liked the cookies you baked.

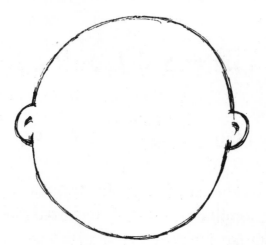

Share Your Favorite Recipe!
Please share a favorite family recipe on the back of this paper. Look for a recipe book filled with class favorites coming home soon with your child!

Name _____ Date _____

Measurement Kit

Dear Family,
It's your child's turn to use the class measurement kit at home!
In this bag you should find:

⊚ a ruler

⊚ a tape measure

⊚ a measurement record sheet

Using the ruler and tape measure, please help your child answer the questions on the measurement record sheet. Assist your child in choosing the correct measurement tool, measuring accurately, reading the measurement, and recording the result on the record sheet. The data your child gathers will be used in a class activity.

You will notice that the first six items to be measured are common objects found around home. Your child will measure different attributes, such as length, height, and width. Discuss these measurements with your child, asking questions such as "Is this wider than . . . ?" or "Are you taller or shorter than this?"

The last four items to be measured are for you and your child to choose. Have fun picking some interesting items. (Fill them in on the record sheet.) Enjoy your measurement adventure together. Please return these materials along with the completed record sheet within the next three days. Thanks for your help!

Sincerely,

The Great Big Idea Book: Math © 2009, Scholastic Teaching Resources

Name _____ Date _____

Measurement Record Sheet

Object	Measurement
1. door	width
2. fork	length
3. table	height
4. window	width
5. refrigerator	height
6. your bed	length
7.	
8.	
9.	
10.	

The Great Big Idea Book: Math © 2009, Scholastic Teaching Resources

Time & Money

Second and third graders are always ready and eager to learn about their world. In particular, the things that adults do and use fascinate them. This includes the concepts of time and money. Second and third graders want to know about time. They love to wear watches, keep track of what's happening when, and show that they can tell time! Even if they haven't yet learned to tell time, it's always on their mind: When's recess? How long until lunch? Money is just as much on their minds. Children delight in opportunities to explore money in real-world situations, using real or play money to learn more about its role in the world.

Time falls under the NCTM standard that addresses measurement. The standard states that "the curriculum should include measurement so that students can understand the attributes of time" and that students need to be able to "make and use estimates of measurement" and "make and use measurements in problems and everyday situations." Money falls under the NCTM standard that covers numbers and operations. The standards explain that students should use multiple models to develop understanding of the place-value structure of the base-ten number system and be able to represent and compare whole numbers and decimals. Certainly, money is one of the models that students will enjoy using.

This section is full of engaging activities that support the standards and will enrich your classroom explorations of time and money. Though the activities are organized by the topics of time and then money, you'll find that many of them make interdisciplinary connections. For example, Tissue-Box Time Lines invites children to write about their days as they explore the passage of time in their lives. (See page 148.) Time Around the World makes a geography connection, as children play a time-zone game. (See page 145.) Poetry Money Math combines language arts and math with "Smart," a humorous poem by Shel Silverstein. (See page 161.)

Eat the Clock!

Introduce students to the parts of the clock with this delicious hands-on activity.

You'll need flour tortillas or large round crackers (one per student), carrot or celery sticks (two per child), raisins, mini-chocolate chips, and paper towels or paper plates.

- Arrange the clock-making materials on a table and have children help themselves to the "parts." (Or prepackage the materials in sandwich bags and give one to each child.)

- Ask children to look at the classroom clock. Guide them to notice the clock face, the minute and hour hands, the numbers (count by fives), and the marks between the numbers. Let children guess which food they could use to represent each part.

- Model for children how to make a clock, using the tortilla for the clock face, the raisins for the 5-minute intervals, the chocolate chips for the minute intervals, and the carrot or celery sticks for the minute and hour hands. As you place the raisins and chips on the clock, show students how to start at the top, placing a raisin in the 12 spot, then placing four chips to represent the minutes between the 12 and 1. Follow with a raisin, and ask students what it represents. (*the 1 on the clock face*) Continue, counting minutes as you arrange the materials to create the hours and minutes all the way around.

- Let children make their own clocks, then invite them to "eat the clock!" Call out different times, starting with the correct time—for example, 12:15. Have students move the hands to show the time, then eat the parts of the clock that represent that time. Continue, letting them make different times and eat their way around the clock!

Eileen Delfini
Richboro Elementary
Richboro, Pennsylvania

TIP

Check for food allergies before children eat their clocks.

ENGLISH Language LEARNERS

Help multilingual students make the connection between time-telling words in English and in their own language. Use a different-colored marker to write these words and phrases in students' native languages. Display these next to the English time-telling words and phrases.

It's About Time Word Wall

Kick off your unit on time by brainstorming key vocabulary associated with time.

Ask students to share any words they know that are connected to time (and don't forget the calendar!)—for example, *day, month, year, time, clocks, minutes, hours, seconds, half past, quarter after, o'clock, a.m., p.m., nighttime, daytime, alarm*. Record the words on posterboard precut in alarm clock shapes and display at children's eye level. Keep blank word cards handy so that students can add to the word wall as they learn more.

TIP

Why do we have 60 seconds in a minute and 60 minutes in an hour? The ancient Babylonians divided each hour in a day into 60 equal parts. Later, the Romans used the same system and called each part *par minuta* (a minute). The Romans also divided the minute into 60 equal parts called *par seconda* (a second).

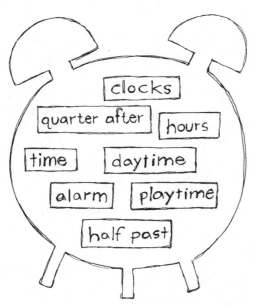

clocks
quarter after
hours
time
daytime
alarm
playtime
half past

Literature LINK

Me Counting Time: From Seconds to Centuries

by Joan Sweeney (Dragonfly, 2001)

This easy-to-read, colorful picture book is perfect for introducing the study of time. It answers the questions: How long is a second? Can you count a minute? What is a decade? How many years are in a century?

The Fives Have It

Getting students to recognize that the numbers on the clock (1–12) represent groups of five can be difficult at first. Help students make the connection between the numbers on the clock and groups of five with the following activities.

- Post index cards next to the numbers telling the minutes on the classroom clock—for example, next to the 1 have the number 5, next to the 2 have the number 10, next to the 3 have the number 15. When practicing telling time from this clock, children can use the index cards to reinforce how the numbers relate to the time.

- Sit in a circle and practice going around the circle, counting by fives to 55. For more fun, toss a beach ball at random to students around the circle, having each child say the next number as he or she catches the ball.

- Play a matching game. Give half the class cards with clock numbers (1–12) written in black and the other half cards with the corresponding minutes (5, 10, 15, 20, 25, and so on.) written in red. Have students find their match.

Eileen Delfini
Richboro Elementary School
Richboro, Pennsylvania

Lakeshore Learning (**www.lakeshorelearning.com**) offers a large assortment of clocks with moveable hands, working cash registers, plastic coins, and other materials for teaching about time and money.

Literature LINK

Clocks and More Clocks

by Pat Hutchins (Aladdin, 1994)

This funny picture book tells the story of Mr. Higgins and his clocks. Mr. Higgins keeps buying more and more clocks to find out which one is telling the correct time. He races around his house from room to room, always finding each clock to be a few minutes behind the other. It's not until a wise clock maker shows Mr. Higgins his errors that he rests.

Time Expressions

Bring language arts into your math lessons by exploring expressions of time.

- Challenge students to listen for expressions about time. For example, children might hear the expressions *time flies; time stands still; there aren't not enough hours in the day; a stitch in time saves nine; the early bird catches the worm; early to bed, early to rise makes a man healthy, wealthy, and wise; time-out; time's a wasting; behind the times; in good time; time on your hands; time of your life; time after time;* and *high time.*

- Invite children to record these expressions in a class learning log and share them with the class. Take time to read them aloud with the class.

- As an extension, have children create an illustrated collaborative banner of their time expressions. Ask each child to choose an expression. Have children write their expressions on sheets of drawing paper and then illustrate them. Arrange the illustrations side by side on a long sheet of roll paper. Glue or tape them in place and display in the hallway for others to enjoy.

What do A.M. and P.M. stand for? Share with students that A.M. is Latin for *ante meridiem*, which means before midday. The initials P.M.. stand for *post meridiem*, which means after midday.

Literature LINK

It's About Time!

selected by Lee Bennett Hopkins (Simon & Schuster, 1993)

Poems by Aileen Fisher, Gwendolyn Brooks, Jack Prelutsky, Charlotte Zolotow, and others fill the pages of this delightful collection of poetry. Charming illustrations capture each hour of a child's day—from waking up "wound" like a clock to thinking bedtime thoughts. Clocks in the upper corner of each page reinforce the concept of passing time.

Make a Timer

With just a few simple materials, students become clock makers to learn about units of time.

- Gather the following materials to make the clocks: plastic liter soda bottles (two per student or team), heavy paper, hole punch, duct tape, and sand or salt.

- Ask students to trace the mouth of one bottle on the paper, then cut out the circle and carefully punch a hole in the center of it.

- Have students pour sand or salt into one bottle, filling it almost to the top. Have them put the paper circle over the mouth of this bottle and place the mouth of the other bottle on top. Have students tape the two bottles together. (If children each make a timer, have them help each other with this step. One child can hold the bottles in place while the other tapes them together.)

- On your signal, have students turn their clocks upside down and time how long it takes the salt to run through to the other jar.

- Let students use their clocks to time each other in fun tasks—for example, how many jumping jacks can they do in the time it takes for the sand to run through?

Bob Krech
Dutch Neck Elementary School
Princeton Junction, New Jersey

Shadow Math

Take advantage of a clear, sunny day to let students turn their shadows into a sundial!

As early in the morning as possible, preferably at the top of the hour, take children outside to a safe, clear, paved area, such as a section of the playground or a school sidewalk. Before proceeding, caution children not to look directly at the sun as this could harm their eyes. Have students line up side by side, and give every other child a piece of sidewalk chalk. Have the children with the chalk trace a neighboring classmate's shadow and then trade places so that each child's shadow gets traced. Have children record their name and the time at the top of their shadows. At the passing of each hour, return to the shadow outlines to trace the top of each student's new shadow and record the time. Invite students to discuss the lengths and times before and after noon. What patterns do they notice? They can record their results using activity page 168.

Experiment with varying the size of the holes students punch in the paper. Have students compare the elapsed time for each. How does the size of the hole change the amount of time it takes for the sand to run through?

Time Around the World

After sharing books about time zones (see Literature Link, below), invite students to explore geography by finding out the time in different parts of the world.

Display a world map and use sticky notes to mark time zones. Start by asking children what time it is somewhere in the world—for example, if you're in New York and it's 10 A.M., what time is it in California? Have a student use the map to determine the answer, then let this child ask the next question. Continue until you've covered all the time zones and plenty of times of day.

Jacqueline Clarke
Cicero Elementary School
Cicero, New York

Find the exact times in cities around the world at Time and Date.com: **www.timeand date.com/world clock**. This site also allows you to check sunrise and sunset times around the world and compare a selected city's time to other time zones!

Literature LINK

Somewhere in the World Right Now

by Stacey Schuett (Dragonfly, 1997)

This picture book is a great way to introduce the concept of time zones. It is full of maps, beautiful illustrations, and easy-to-read, interesting text. It follows activities occurring simultaneously around the world in different time zones. *Nine O'Clock Lullaby*, by Marilyn Singer (Harper Trophy, 1993), offers a more lyrical look at time zones. A rhythmic lullaby lets readers "travel" through the different time zones to see what's going on around the world when it is 9:00 P.M. in New York, including the mid-Atlantic at midnight, England at 2 A.M., and Australia at noon.

Sundial Time-Tellers

Have students see how a real sundial works by building this class model. You'll need a sheet of white tagboard, a nail, and a pen.

At the start of a school day, place a sheet of tagboard on the ground outside in sunny spot. Use a few small rocks to hold the tagboard in place. Poke a long nail through the center of the tagboard. Make sure you leave most of the nail showing. (You might use a small bit of clay at the base of the nail to add stability.) Mark the tip of the nail's shadow with a line, and record the time. Repeat this procedure each hour, marking the tip of the shadow and recording the time. Let children use the sundial the next day to tell time. For a challenge, have them estimate the time when the shadow falls somewhere between the lines they've marked. Compare their estimates with the actual time. Let students try again later. Are their estimates closer this time?

Clock Concentration

A fun way for kids to practice matching digital times with analog clocks is this game of Concentration.

Make one copy of the reproducible cards on pages 169 and 170. Complete the cards by filling in matching times on the digital and analog clock cards, then make multiple copies of each. Divide the class into small groups and give each a set of cards (one from each page). Have children cut apart the cards, mix them up, and place them facedown. Have players take turns turning over two cards to see if they match. If a player gets matching analog and digital clock cards, he or she takes the cards. If not, the cards are turned facedown again and the next player takes a turn. Play until all cards have been matched.

TIP

Help introduce the concept of the sun's position in the sky, shadows, and how sundials work, with "Shadow Race," a poem by Shel Silverstein from *A Light in the Attic* (HarperCollins, 2005).

Kid Clocks

This lively activity invites students to become the parts of a clock as they practice telling time.

- Review the parts of the clock (numbers, hands, circle shape). Write the numbers 1 to 12 on large index cards (one number per card) and give one to each of 12 children.

- Have these children arrange themselves in a large circle. Invite two volunteers to be the hour and minute hands of the clock. Have one lie inside the circle with feet placed at the center of the circle and hands stretched out over his or her head to be the minute hand. Have the other lie inside the circle (again with feet at the center of the circle) with arms at the side to be the hour hand.

- Let the remaining students take turns calling out a time and helping the "clock hands" get into position. After changing the time several times, let students trade places so that they can all be a clock part.

- Turn the clock number index cards over and write the corresponding Roman numerals. Use the cards to make a "kid clock" with a Roman numeral clock face.

Rock Around the Clock

Help children learn about elapsed time with an activity that combines math, music, and movement. You'll need a clock with moveable hands, a CD or tape player, and some lively music.

Set the clock to a particular time. Have students record the time on a sheet of paper. Invite a student to stand up in front of the class and hold the clock so that everyone can see it. Start the music while the volunteer slowly pushes the hands around the clock. Kids may move around or dance in their places while the music plays. When you stop the music, have the child holding the clock stop moving the hands. Have the other children sit down. Ask students to figure out how much time has passed on the clock. After discussing the results, repeat the activity, letting another student hold the clock and move the hands.

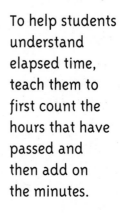

TIP

To help students understand elapsed time, teach them to first count the hours that have passed and then add on the minutes.

Tissue-Box Time Lines

Here's an easy way for students to create a three-dimensional time line about their day.

◉ Give each student white construction paper, markers or crayons, and a square tissue box. Have students trace one of the upright sides of the tissue box onto the construction paper, cut it out, then use it as a template to cut out three additional squares.

◉ Ask students to think about four different activities they do in one day—for example, get on the bus, have recess, eat lunch, and go to bed. Invite them to write about and illustrate each activity on the four squares of paper, then record the time of the day they usually do the activity.

◉ Have students glue the four papers in order to the sides of the box. Give students time to compare their activities and the times at which they do them.

A.M.–P.M. Flip Book

Make these fun books to strengthen students' understanding of the differences between A.M. and P.M.

◉ Have students fold a sheet of paper in half (vertically), then turn the paper horizontally (with the fold at the top) and cut two slits, as shown, to divide the front flap into thirds.

◉ Ask students to choose a favorite morning activity and to illustrate it on the front of the first flap. Have them write the time (including A.M.) on the flap. Have students repeat this procedure to complete the front of the other two flaps—illustrating a favorite morning activity and recording the A.M. time.

◉ Have students do the same under each flap, this time illustrating three favorite afternoon or evening activities. Have them record the time for each, being sure to add P.M.

◉ Invite students to share their flip books with one another, comparing favorite activities and times that they do them.

Peg Arcadi
Homeschool Teacher
Trumansburg, New York

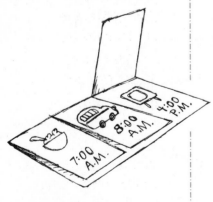

How Long Will It Take?

How long does it really take a minute to pass by? Discuss with students how long a minute is (60 seconds). Then try this activity to reinforce the concept of time and strengthen subtraction skills, too.

Have children count slowly to 60 to get a feel for how long 60 seconds lasts. Next, ask students how many things they think they can do in a minute—for example, how many times they can write their names, snap their fingers, hop, do jumping jacks, tie their sneakers, and say the alphabet. List ideas on the board. Give each child a copy of page 171. In the first box, have students record an estimate of what they can do in one minute. Pair up students and give them a stopwatch. While one partner tries the activity, have the other partner time a minute on the stopwatch. After completing the activity, ask students to calculate the differences between what they thought they could do in a minute and what they actually did. Have students use the remaining boxes to estimate and try out two additional one-minute activities.

Wendy Wise-Borg
Maurice Hawk Elementary School
Princeton Junction, New Jersey

Literature LINK

Dear Benjamin Banneker

by Andrea Davis Pinkney (Voyager, 1998)

This beautifully illustrated picture-book biography about Benjamin Banneker is a great introduction to the first African American to write and publish an almanac. His almanacs were full of accurate calendars. He also was famous in his day for making a wooden striking clock that kept accurate time for more than 50 years—one of the first and finest wooden clocks made in Colonial America! This history maker was also an astronomer, a farmer, a surveyor, and a mathematician!

TIP

To obtain an informal assessment of students' understanding, post a time that includes the notation of A.M. or P.M. Ask students to copy the time on a sheet of paper and draw a picture of an activity they might perform at that time.

Calendar Clues

Invite your students to become number detectives while trying to figure out a secret date on the calendar.

Before starting, come up with a set of number clues that correspond to dates on the calendar—for example, "This number is equal to half a dozen" or "This number is half the number of students in our class." Hide the clues under dates on the classroom calendar, being sure to place them so that one clue will lead to the next. (You can use sticky notes to make flaps on the calendar squares. Write the clues under the flaps.) Under the last flap, write a surprise that students will get—for example, an extra recess or free-reading time. Come up with a separate clue that will lead students to the first flap—for example, "This is the number of inches in a foot." (This clue will lead students to the number 12, under which they'll find the next clue.) Let students take turns solving clues until they get to the secret date, where they'll find their surprise!

Save That Date

Invite students to explore each month of the year by having them talk with family members at home about important dates to remember.

Make copies of Save That Date, a reproducible monthly calendar. (See page 172.) Have students take the calendar home and discuss with their families dates that are important in each month—for example, family birthdays, holidays, and so on. Ask students to write the dates in the boxes and draw a picture that symbolizes the activity, person, holiday, or event. that makes this date important.

When is International Joke Day? How about Amelia Earhart's birthday or National Dog Week? Find out when these and other dates fall at Surfing the Net with Kids: **www. surfnetkids.com /sept.htm**, where you'll find listings and facts for just about every day of the year!

Round and Round Clock Rhyme

Use the following rhyme during transition times or as a warm-up activity to reinforce telling time.

Have students get out their notebooks, or supply them with lined paper. Using a clock with moveable hands, recite the following rhyme while moving the hands around and around. Stop the hands when you get to the end of the rhyme.

> **Round and round and round it goes,**
> **Where it stops, nobody knows,**
> **Where it stops will be the time,**
> **Write the time on the very next line!**

Have children record the time shown on the clock, being sure to add A.M. or P.M. if this is a focus of your time-telling lessons. This is a great way to keep kids engaged while doing rote practice. When students get the hang of it, let them take turns moving the clock hands as they lead the class in the rhyme.

Barbara Gauker
Immaculate Conception School
Bristol, Pennsylvania

Invite your students to travel back in time to learn about timekeeping practices of the past. A Walk Through Time (**physics.nist.gov /GenInt/Time/ time.html**) features ancient calendars and early clocks and includes several examples of interesting and strange time-keeping devices.

The Grouchy Ladybug

by Eric Carle (HarperCollins, 1996)

The colorful die-cut pages of this classic picture book follow the story of a grouchy ladybug as she picks a fight with increasingly larger animals. A clock face at the top of each page lets children track the passing of time as the suspense in the story builds. As a follow-up, let children write a collaborative story about their day, using a clock face on each page to show the passing of time.

Comparing Clocks

Students are so often used to seeing digital clocks in their world that it can be extra-confusing for them to learn how to tell time with analog clocks. Use a Venn diagram to help them see similarities and differences between the two types of clocks.

Divide the class into groups. Give some groups a working analog clock and others a working digital clock. Give groups with the analog clocks circle-shaped cards, and those with digital clocks rectangle-shaped cards. Ask students to examine their clocks and to record their features on the index cards (one feature per card). Bring students together to share the cards. Make a Venn diagram to learn more. Draw two large overlapping circles. Label one circle "Analog" and one "Digital." Let children take turns placing their cards in the appropriate circles, indicating whether the features apply to one or both clocks. Discuss the results: How are the clocks alike? How are they different?

Time Check!

This activity strengthens students' time-telling abilities and provides a quick and easy informal assessment.

Give each student a copy of the record sheets on page 173. Have children tape one of the record sheets to their desks. Model the activity by announcing that it's time for a "time check." Check the clock, tell the time, and write it on the board. Have children write the time on their papers, too. At selected times throughout the school day, ring a bell or announce that it is a "time check" moment. Have students stop what they're doing, look at the classroom analog clock, and record the time. Keep a master for your records. Collect the sheets at the end of the day or when all the spaces have been filled. Check students' records against your own for assessment purposes. As students' abilities grow, ask questions about elapsed time—for example, "How many more minutes until we go to art? How long has it been since the last time check?"

Stacey Brandes
St. Joseph School
Burlington, Vermont

TIP

Invite students to bring in and share extra clocks from home, including alarm clocks, digital clocks, clocks with Roman numerals, clocks without numbers, watches, and so on.

Stories Tell Time

Using literature to teach math concepts helps students make connections, enjoy literature, and view math in a different way.

Divide the class into groups. Give each group a book that relates to time in some way—for example, *The Very Hungry Caterpillar,* by Eric Carle, is a classic for teaching days of the week. Other concepts to cover include hours, months, seasons, elapsed time, and schedules. Invite students in each group to share the book among themselves and then brainstorm ways the book teaches about time. Let each group prepare a book-based mini-lesson about that concept. (Model a mini-lesson first and discuss lesson formats students might use.) These mini-lessons lead to lots of creative thinking, and students love playing the part of teacher.

Bobbie Williams
Brookwood Elementary School
Snellville, Georgia

Take-Home Activity:
Family Planner

With today's busy families, a planner is always helpful for organizing family events and activities. Help students practice using time schedules as they work together with their families to plan out a week of activities and events.

Make copies of the reproducible daily planner on page 174. Have students take home the planner and complete it with their families for the following week. Families can post the planner on refrigerators or another handy spot and refer to it as they plan their time. Ask students to bring the planners back to school after the week is over. Discuss similarities and differences in schedules—for example, How many students noted doctor's appointments on their schedules? Soccer practice? Meetings? Special occasions?

Make a Theme Clock

Celebrate your study of time by having children design clocks that connect with their own interests and hobbies.

- ⟳ Share with students magazine and catalog pictures of unique clocks and watches. Discuss the various hobbies and interests the clocks might represent—for example, a bird lover might appreciate a clock with pictures of birds that also chirps their songs on the hour.

- ⟳ Have students write down a couple of interests or hobbies on a sheet of paper. Under each topic, have students make a list of picture ideas that match their topic—for example, balls that represent favorite sports, or different kinds of pets for an animal lover.

- ⟳ Have students use the reproducible clock face (see page 175) to create a clock with their hobby or interest as a theme. Make a checklist on the board to remind students of the features, including hands, 5-minute symbols, and marks for minutes.

- ⟳ Invite students to arrange their clocks on a bulletin board and to come up with a name for their display.

Literature LINK

A Quarter From the Tooth Fairy

by Caren Holtzman (Scholastic, 1995)

This easy reader tells the story of a boy who receives a quarter from the Tooth Fairy. He thinks about all the things he can buy for twenty-five cents, but gets so confused that he decides to buy his tooth back.

Money Word Wall

Kick off a study of money by building a word wall that will help you assess what students already know and serve as a teaching tool for more learning.

Invite students to call out words associated with money—for example, *cash, dollar, cents, value, bank, pay, save, hundred, quarter, nickel, coins, pennies, dimes, nickels, half-dollars*. Write the words on tagboard shapes that represent money (for example, rectangular dollar-bill shapes and circular coin shapes). Display the word cards at children's eye level. Keep blank word cards handy so that students can add to the word wall as they learn more about money. Use the word wall as a source of skill-building games—for example:

- Strengthen spelling by starting to spell a "mystery word" on the word wall. Say each letter slowly, letting children take turns spelling the rest of the word as soon as they discover which one it is.

- Build concepts by letting children sort the word cards. (If you attach them with Velcro®-type hook-and-loop fasteners, they'll be easy to put up and take down.) You might start a sorting pattern—for example, grouping cards by words that name values (*quarter, dollar,* and so on) and words that don't—then let children come up one at a time to add a word to a group.

- Let students take turns reading the words in alphabetical order. Use words that start with the same letter—for example, *cash* and *cents*—as the focus of a mini-lesson to learn more about alphabetizing.

TIP

Someone who collects coins is called a numismatist. Conduct a class survey to find out what students collect.

ENGLISH Language LEARNERS

Strengthen students' understanding of the word *change* by discussing the multiple meanings of the word. Show students the word as a verb in the sentence: "Can you change a 5-dollar bill for 5 ones?" Show students the word as a noun in the sentence "The change in my pocket is gone!" Show students the word as another noun in the sentence "How much change do I get?" Let students brainstorm other words with more than one usage—for example, *saw, present,* and *can.*

TIP

Some of
the world's
currencies:
Burundi: franc
Cambodia: riel
China: yuan
Germany: euro
Great Britain:
   pound
India: rupee
Japan: yen
Mexico: peso
Nicaragua:
   córdoba

Money Around the World

Introduce students to currencies around the world with this fun fact: The smallest money ever used was the Greek *obelas*. It was smaller than an apple seed. People kept the obelas in their mouths, hoping they wouldn't sneeze or swallow! Learn more about money around the world with an activity that integrates math with geography, literature, and the Internet.

- Invite students to bring in from home any coins their families might have from other countries. Let students share their coins and the names for them if they know them. Use the information on this page (see Tip, left) to introduce the names of other currencies around the world.

- After learning the names of some of the different currencies, share *Money,* by Joe Cribb (Dorling Kindersley, 2000), a book of photos and fun facts about the history of money. Encourage students to look for photos that correspond to any of the coins they brought in. Share interesting information about these coins.

- For more fun, visit Web sites that show currency conversions. Challenge children to find out what it would cost to buy, for example, a school lunch in other countries. Start with countries represented by coins children shared from home. Two Web sites to try are 164 Currency Converter by OANDA (www.oanda.com/cgi-bin/ncc) and Universal Currency Converter (www.xe.net/ucc).

ENGLISH Language LEARNERS

Use a lesson on world currency to build bridges between your English language learners' countries of origin and what they're learning in class. Invite children to create a poster that features their country's currency. Have them include words for coins, bills, and so on. Display the poster next to one that shows the currency they're learning about. Make comparisons: Are the coins a similar size? How do their values differ?

Let's Make Change

Students practice making change with this musical role-playing activity.

🎵 Divide the class into pairs. Assign students the role of buyer or seller. Supply the sellers with resealable bags filled with $3.00 worth of plastic coins. Supply the buyers with resealable bags containing several green sheets of construction paper cut to the size of dollar bills.

🎵 Invite the sellers to choose three items (for example, pencil, marker, and eraser) from their desks that they would like to sell. Have them use sticky notes (with their names) to put a price on each item (less than one dollar). Then have sellers display the items on their desks.

🎵 When sellers are ready, have buyers line up in front of their desks. Play some music and have the buyers move from one desk to the next. When you stop the music, buyers stop and purchase something at the nearest desk. The seller must give the buyer correct change.

🎵 Have students switch roles and money bags after taking several turns. Remind students that all items will be returned after the activity.

Becky Mandia
Newtown Elementary
Newtown, Pennsylvania

Literature LINK

My Rows and Piles of Coins

by Tololwa M. Mollel (Clarion, 1999)

This award-winning picture book tells the story of an African boy named Saruni. Saruni dreams of buying a bicycle and saves all his coins for months and months, but he discovers he does not have enough. Saruni is assisted by his father and in return begins saving to help his family by a cart. An author's note at the end describes some of the African coins featured in the story and their values in relation to other coins.

TIP

Strengthen students' coin-counting abilities with the software program *Coin Critters* (Nordic Software). Students practice distinguishing between U.S. coins, matching coins, and counting change.

Book Orders Add Up

Classroom book orders provide a great way for students to practice using money in real-life situations. Use your book order forms with the following activities:

- Have students select several books they would like to have, record them on the order form, and use paper and pencil or a calculator to add up their purchases. (Make additional copies so that students can use the forms for actual orders as well as for fun and practice.)
- Give students a set amount they can "spend." Challenge them to see who can get the most books for that amount of money.
- Have students put the books in order from highest to lowest price.
- Create word problems using students' names—for example, "Mick and Lucy each have three dollars. They want to combine their money to buy and share books. What are some book combinations they can buy?"
- Incorporate literature by inviting students to add up the number of books in the same genre, on the same subject, by the same author, and so on.

Share with students the fact that paper money was first issued in the United States on March 10, 1862, and became legal tender by an act of Congress seven days later.

Literature LINK

Benny's Pennies

by Pat Brisson (Yearling, 1995)

A boy named Benny has five new pennies—and five different ways to spend them. Will he be able to buy all five things? This cumulative story is a great read-aloud and teaches a lesson not only about the value of pennies but also about spending wisely.

Money Makes History

Why are certain people from history honored by having their face on United States currency? Encourage students to find out by learning about the lives of people depicted on U.S. coins and bills.

I nvite students to team up to research someone pictured on a coin or bill. Encourage students to find out why they think that person was chosen. Challenge students to present their information to the class in a novel way: Have students create a poster-size illustration of the coin or bill they researched. Show them how to cut out the center of the coin or bill to make room for their heads to peek through. Let students present their biographical information in the first person—for example, "I am Sacagawea and I am pictured on the new dollar. I helped Lewis and Clark on their expedition out west." Children in each group can take a turn being the face on the coin or bill and sharing information.

Literature LINK

Alexander, Who Used to Be Rich Last Sunday

by Judith Viorst (Aladdin, 1980)

This classic picture book describes the story of Alexander, who received a dollar from his grandparents one weekend. Little by little the money disappears as he spends some on bubble gum, rents a snake for an hour, and gets fined by his dad, among other things. Invite your students to interact with the story as you read it aloud. Supply each student with seven dimes, four nickels, and ten pennies. As students hear about Alexander losing a specific amount of money, have them take away the same amount from their piles. For a challenge, invite students to count and write the amount of money Alexander has at different stages of the story.

TIP

Learn about the history of money at **www.wdfi.org /ymm/kids/ history/default. asp**. You'll also find quizzes, pictures of money, and links to related sites.

Design a Coin

Children learn about the features of coins by designing their own!

Make copies of the reproducible coin templates on page 176. Review the parts of a coin by having students take a close look at real coins. What do they have in common? How are they different? Let students sketch their coin designs first, determining the value, symbols, color, and words to appear on the front and back. They can transfer their designs to the templates, cut out the coins, and glue the front and back together (back sides together). Let students write about the features of their coins on a separate sheet of paper. Do their coins feature a person or place? Does the coin represent a value other than that of an existing coin? Display the coins by punching a hole through the top and pulling string through to hang from the classroom ceiling. Display the descriptions on a bulletin board so that readers can look up to find the matching coins.

State by State

Get your students excited about United States geography and money at the same time.

Gather as many different state quarters as you can and place them in a jar. Organize students into groups of three or four. Invite each group to pick a quarter from the jar. Have students research their group's quarter to find out more about the symbolism of the picture on the back and learn some fun and interesting facts about that state. Have each group create a presentation about the coin and the state it represents.

Mitzi Fehl
Poquoson Primary School
Poquoson, Virginia

TIP

Visit the Official United States Mint Web site (**www.usmint. gov**) for information about the U.S. State Quarter Program. For states that have not released their quarters or announced their symbol, invite students to research the state and select a fitting state symbol.

Poetry Money Math

Share a thought-provoking poem to help children understand that more coins don't always equal more money. Then play a fast-paced game to learn more.

- Read aloud Shel Silverstein's poem "Smart," from *Where the Sidewalk Ends*. (See Literature Link, below.) As they listen, ask children to think about each trade the boy makes. For example, when he trades a dollar bill for two shiny quarters, is he making a "smart" trade? Why?

- Let children share their thoughts about each trade in the poem. Then have them team up with partners to write new verses. Reread the poem, substituting students' lines for original verses. Again, discuss whether or not the trades are "smart."

- Follow up by letting students play Rolling for Coins, a game that reinforces concepts about money. Give pairs of children two copies of the record sheet (see page 177), a die, and a bag of assorted coins. Have children take turns rolling the die and taking the corresponding number of coins from the bag (without looking). Have children count up the value of the coins and record both the number of coins and their value on the record sheet. For each round, have students circle the highest number rolled and put a square around the highest value.

- After letting children play several rounds, bring them together for a discussion. Does the highest number rolled always result in more coins? (*Yes*) Does this always result in more money? (*No*)

Literature
LINK

Where the Sidewalk Ends

by Shel Silverstein (Harper & Row, 1974)

What would your students say if someone offered to trade them two shiny dimes for one quarter? Share the poem "Smart" to challenge your students' thinking through problems like this. In the poem, a child thinks he is really smart because he trades one quarter for two shiny dimes (because two is more than one), two dimes for three nickels (because three is more than two), and so on.

The software program *Dollarville* (Waypoint Software) is divided into four levels of difficulty and provides students with opportunities to learn all about money. A character named Deputy Dollar guides students through the game.

Coupon Clippers

Here's an easy game to play that uses coupons to strengthen students' skills in comparing different amounts of money.

- Collect newspaper coupon circulars. Have students bring some in from home, too. Cut out the coupons as they come in. You'll need about 20 per child.

- When you have enough coupons, divide the class into pairs and give each child a set of coupons. Have children shuffle their coupons and stack them facedown.

- To play, have both players place the coupon on top faceup at the same time. The player with the higher money amount on his or her coupon takes both. If the value of the coupons turned over is the same, have children turn over another coupon, continuing until one is higher than the other. The player with the higher value takes all the coupons from that round.

- Play continues until all coupons have been turned over and taken. The player with the most coupons wins. As a follow-up, invite students to work together to arrange the coupons in order from least value to greatest value.

Literature LINK

Coin County: A Bank in a Book

by Jim Talbot (Innovative Kids, 1999)

Get children excited about saving money with this interactive picture book. It's told in rhyme and covers the many ways that coins add up to a dollar. Children can save more than $20.00 by filling the coin slots built into each page. With slots for pennies, nickels, dimes, and quarters, students can practice counting with real money. (This might make a fun class project; students can choose a charity to donate their coins to when they've filled all of the slots.)

TIP

Visit Ron's Currency, Stocks, and Bonds at **www.ronscur rency.com/rhist. htm** to see a time line of the history of paper money in the United States. Find out answers to the following questions: Which bank was the first to issue paper money? When did "In God We Trust" first appear on paper money?

Counting Coins Clapping

When counting coins, children must be able to count by ones, fives, tens, and twenty-fives. Let them practice counting by these numbers with an activity that engages their kinesthetic learning styles.

- Gather enough real or plastic quarters, dimes, and nickels to count to 50 with each (two quarters, five dimes, and ten nickels).

- Call out the number 50. As you drop two quarters in a container, have students call out the value and clap the number of quarters. "25, 50!" (clap twice)

- Repeat the procedure for dimes, having students count and clap each group of ten. Now try nickels. Each time, help students make the connection between smaller-value coins and larger-value coins.

Literature LINK

The Coin Counting Book

by Rozanne Lanczak Williams (Charlesbridge, 2001)

"One penny, two pennies, three pennies, four. What will we get when we add just one more?" That's the way this book of verse begins. Full of color photos of coins, this book introduces the counting of coins all the way up to the Sacagawea dollar!

TIP

When students are well practiced in counting by these numbers, try mixing it up. Count to 50 using dimes and nickels. (Have them watch closely to see which coin you are dropping into the container.) Add some pennies to the combinations, too.

Pocket Change

Get your students excited about counting change and familiarize them with coin characteristics at the same time.

- Give each child a copy of the reproducible activity sheet on page 178, some coins, and a crayon.

- Ask children to hide some of the coins in the pocket (placing the coins under the paper and making sure they are inside the outline of the pocket).

- Have children trade seats, leaving their papers and hidden coins on the desks. Once children are seated, ask them to estimate the amount of change in the pocket by feeling the coins through the paper. Then have them record their estimates in the space provided, being careful not to move the coins.

- Invite children to use the crayons to make a colorful rubbing of the hidden coins. Ask them to revise their estimates if they wish (based on the rubbings). Then have them lift the paper to reveal the hidden coins and record the actual amount of change in the pocket. Use the rubbings to create a colorful display that reinforces coin values and characteristics.

Coins Weigh In

Strengthen familiarity with coins and counting skills with an activity that invites children to measure weights.

Set up a learning station with several balances and containers of quarters, dimes, nickels, and pennies. To model the activity, ask students to predict how many dimes it will take to balance the scale with one penny. Try it out and discuss the results, reviewing both the number of dimes it took and the value—for example, it took two dimes, or 20 cents worth of dimes, to equal one penny. Set up a chart to record information, providing space to record predictions and results. Let students visit the station in small groups to test out more combinations, recording their predictions and results on the chart.

When teaching children how to count change, remind them to first group like coins together and then arrange them from greatest to least value.

Money Match

Turn coin-counting practice into a game with this easy idea.

Let students pair up with a classmate to play this game. Give each player a set of plastic coins that includes multiple half-dollars, quarters, dimes, nickels, and pennies. Have players take turns choosing and displaying an amount of money under $1.00. The other player needs to match the amount of money, using a different combination of coins.

Make a Date With Money

Add a pink construction paper piggy bank to your class calendar to set up for some practice with coins and bills.

Place real or plastic money (including pennies, nickels, dimes, quarters, half-dollars, and dollar bills) in clear contact paper squares, leaving a space at the top for punching a hole. Stack the coins by value, and use straight pins to hang on a bulletin board with the class calendar. Each day, invite a child to select money with a value equal to the day's date—for example, if it's the third day of the month, the child will select three pennies. Have the child use straight pins to attach the money to the piggy bank. When the date gets to the fifth of the month, the pennies have to be exchanged for a nickel, and so on, up to the dollar bill. Students love doing this and it teaches a lot about the value of the coins and provides practice with exchanging coins.

Judy Meagher
Student Teacher Supervisor
Bozeman, Montana

Use the book *Math for Smarty Pants,* by Marilyn Burns (Little, Brown, 1982), to enrich your lessons on money. Activities include Money Alphabet, in which a monetary value is assigned to each letter of the alphabet. Children will love exploring the value of their names.

Calculating Classmates

Students will have fun combining and adding coins while playing this game with their classmates.

Obtain oversize coins from a teacher supply store or create them yourself. Attach string to each coin to make a necklace. Give each student a coin necklace. Play music while students walk around the classroom. When the music stops, have each student find a partner and sit down. Ask partners to calculate their totals. Have partners share their totals with the class to check for accuracy. Play again, repeating the procedure but making the game more challenging by giving each child more than one necklace. Or, when the music stops, have children form groups of three or four, then calculate their totals.

Literature LINK

A Dollar for Penny

by Dr. Julie Glass (Random House, 2000)

In this rhyming easy reader, a girl named Penny opens a lemonade stand, and watches her earnings add up to a dollar. A table of coins and values appears at the back.

ENGLISH Language LEARNERS

Make a money reference tool for English language learners by creating a chart with three columns labeled "Coin," "Name," and "Value." Give children pictures of coins (or plastic coins) to glue in the first column of the chart. Have them complete the chart by recording the names of the coins and their values. Let them attach the charts to their desks for easy reference.

Take-Home Activity:
Counting and Comparing Coins

The riddles on page 180 let families work together to learn more about money.

Give each child a copy of page 180. Review the page, making sure children know that these are riddles they can solve with a family member. Have children notice that there are three riddles on the page, plus space for creating a new riddle to share with the class. When children return their papers to class, compile their new riddles. Use them to create new take-home activity pages for reinforcing money skills and concepts!

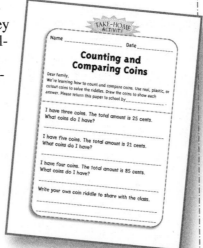

Guess My Coin Combination

Children always enjoy a good guessing game. This fun and challenging game for partners will strengthen students' skills in using different combinations of coins to count to a specific amount.

Students will need sticky notes, pencils, and an assortment of real or plastic quarters, dimes, nickels, and pennies. The first player writes an amount that is less than $1.00 and records a combination of coins that equals that amount, then covers the amount and combination with a sticky note. The first player tells the second player the hidden amount and how many coins make up the secret coin combination. The second player tries to guess and show the correct combination.

When teaching children to count coins, remind them to group like coins together. Try to start with the coin with the highest value, then the next highest, and so on. This comes in handy for purchasing something with the fewest number of coins.

Name _____

ACTIVITY PAGE

Shadow Math

Activity: _____

Activity: _____

Activity: _____

Activity: _____

Name _____ Date _____

Clock Concentration Cards

Name _____   Date _____

Clock Concentration Cards

: _ _ _ _ _ _ _	: _ _ _ _ _ _ _
: _ _ _ _ _ _ _	: _ _ _ _ _ _ _
: _ _ _ _ _ _ _	: _ _ _ _ _ _ _
: _ _ _ _ _ _ _	: _ _ _ _ _ _ _
: _ _ _ _ _ _ _	: _ _ _ _ _ _ _

Name _____ Date _____

How Long Will It Take?

Activity: _____

Estimated

Actual

Activity: _____

Estimated

Actual

Activity: _____

Estimated

Actual

Name _____   Date _____

Save That Date

Practice reading a calendar with family members, and find out some dates that are important enough to save.

January	February	March
April	**May**	**June**
July	**August**	**September**
October	**November**	**December**

The Great Big Idea Book: Math © 2009, Scholastic Teaching Resources

Name _____   Date _____

Time Check!

Time Check	Time Check	Time Check	Time Check
1. _____	1. _____	1. _____	1. _____
2. _____	2. _____	2. _____	2. _____
3. _____	3. _____	3. _____	3. _____
4. _____	4. _____	4. _____	4. _____
5. _____	5. _____	5. _____	5. _____
6. _____	6. _____	6. _____	6. _____
7. _____	7. _____	7. _____	7. _____
8. _____	8. _____	8. _____	8. _____
9. _____	9. _____	9. _____	9. _____
10. _____	10. _____	10. _____	10. _____

TAKE-HOME ACTIVITY

Family Planner

Dear Family,

We are learning about time in school. One concept we are studying is using time to make schedules. Please help your child make a schedule of your family's important activities for the week.

Day: Date:	Day: Date:	Day: Date:	Day: Date:	Day: Date:	Day: Date:	Day: Date:
Activity:	Activity:	Activity:	Activity:	Activity:	Activity:	Activity:
Time:	Time:	Time:	Time:	Time:	Time:	Time:
Activity:	Activity:	Activity:	Activity:	Activity:	Activity:	Activity:
Time:	Time:	Time:	Time:	Time:	Time:	Time:

Name _____   Date _____

Make a Theme Clock

Name _____   Date _____

Design a Coin

The Great Big Idea Book: Math © 2009, Scholastic Teaching Resources

Name _____   Date _____

Rolling for Coins

Player 1		Player 2	
Number of Coins	Value	Number of Coins	Value

The Great Big Idea Book: Math © 2009, Scholastic Teaching Resources

Name _____ Date _____

Pocket Change

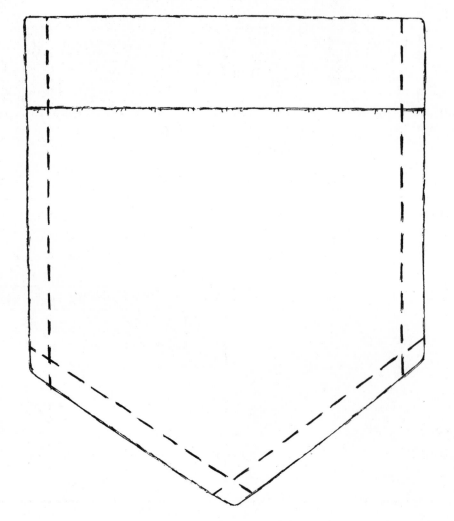

Estimate 1: _____   Estimate 2: _____

Actual Amount: _____

Name _____   Date _____

Coin Chart

	Penny	Nickel	Dime	Quarter
Value				
Picture on Head				
Picture on Tail				
Number in One Dollar				
Color				
Other Details				

Name _____   Date _____

Counting and Comparing Coins

Dear Family,
We're learning how to count and compare coins. Use real, plastic, or cutout coins to solve the riddles. Draw the coins to show each answer. Please return this paper to school by _____ .

I have three coins. The total amount is 25 cents.
What coins do I have?

_____

I have five coins. The total amount is 21 cents.
What coins do I have?

_____

I have four coins. The total amount is 85 cents.
What coins do I have?

_____

Write your own coin riddle to share with the class.

_____

The Great Big Idea Book: Math © 2009, Scholastic Teaching Resources

Graphing

Kids are born collectors! At a very young age, they acquire great numbers of rocks, trading cards, action figures, fast-food-meal toys, and a vast array of other objects. They repeatedly sort, count, and re-count their collections and love to watch the numbers grow as they add new pieces.

In school, we can build on these first experiences with graphing to help students make sense of the information-rich society in which they live, and to help them become informed citizens and intelligent consumers. We know they will be easily drawn into the process of collecting data, but we must be equally sure that they know how to organize and present the information they gather.

According to the NCTM *Principles and Standards* (2000), students need to learn to "formulate questions that can be addressed with data . . . collect, organize, and display relevant data to answer them . . . develop and evaluate inferences and predictions that are based on data." This section is filled with activities that teach students how to collect, organize, and describe data. The activities are drawn from all areas of the curriculum and provide opportunities for students to work individually, in small groups, and as a class. As you flip through this section, you'll see activities to teach line, bar, circle, and picture graphs, as well as other ways to organize data—for example, by using glyphs, coordinate grids, and Venn diagrams.

Story Time Graph

Help students see how they can use graphs to make shared decisions, such as what book to read during story time.

Choose two books and display them on the ledge of the board. Invite students to preview them during snack time or recess. Give each child a clothespin. Ask children to vote for the book they would like to read during story time by clipping the clothespin to the cover. When everyone has had a chance to vote, gather children around the books and ask the following questions:

- What do the clothespins represent?
- How many students wanted to hear _____ for story time? (Repeat with the other title.)
- How many more students wanted to read _____ than _____?
- Which book will we read today? Why?

Use graphs to involve students in other classroom decisions such as what topics to study, where to go for field trips, or when to have recess.

Photo Faces

Be prepared for your next graphing experience with these reusable, student-created graph markers.

Take a head shot of each child (or have children bring in photos from home). Give each child half an index card. Have students glue their photos to the center of the card, and create a decorative border around the edges. Laminate the index card graph markers for durability. When it's time to graph, have students attach their personalized markers to the surface using Velcro®, masking tape, or removable wall adhesive. A quick glance at the remaining markers will tell you which students have missed their turn!

Kelley Foster
Cicero Elementary School
Cicero, New York

Let's Talk Graphing Word Wall

Encourage your students to "talk math" by creating a word wall of graphing terms.

- Ask students to name words that relate to the topic of graphing. List each word on a separate index card.

- Arrange the words in a brick formation on a bulletin board. Label the board, "Math Is Spoken Here!"

- As you continue to study graphing, add new "bricks" to the wall.

- Use the wall to play word games such as Pictionary. Let students take turns choosing a word and drawing it on the board for classmates to guess. The first child to name the word correctly takes the next turn.

Marianne Chang
Schilling School
Newark, California

ENGLISH Language LEARNERS

Work with English language learners to create their own picture dictionary of graphing terms. On each page of a student-made book, record one of the words from the wall. Invite children to illustrate their books to provide picture clues for math vocabulary.

Look What I'm Reading!

Are your students reading from different genres? Use graphs to track their preferences.

Make one copy of the bar graph for each student. (See page 210.) Review the different genres listed on the horizontal axis. Share a sample book for each and invite children to add titles they've read that fit in the various categories. (Poetry, Informational, Science Fiction/Fantasy, Mystery, Biography, Folklore, Realistic Fiction, Historical Fiction)

Instruct students to record the titles of books they read in the appropriate columns on their own graphs. Meet with students on a regular basis to review their graphs. In which genres are they reading the most books? the fewest? Use this information to encourage them to explore other genres.

Add these graphs to students' language arts portfolios. Let children use them at open school night or conferences to show their families what types of books they're reading.

Graph a Story

Use a line graph to plot the intensity of main events and show students how stories develop.

Work with children to record the main events of a story in chronological order on a sheet of chart paper. Assign each event a number and record these numbers on the horizontal axis of a line graph. Label the axis "Story Events." Number the vertical axis from 1–10 and label it "Intensity Scale." Rate each event on the line graph using the intensity scale. Ask students to look at the graph and find the climax of the story. Discuss the position of the climax in relation to other events. Explain that authors often use this pattern to tell a story. Invite students to use line graphs to plot the events of other books they are reading.

Graphing Freewrite

"Freewriting" is a timed exercise in which students record their thoughts continuously about a specific subject without regard for grammar or spelling. Here's how you can apply this technique to reading and interpreting graphs.

Use the overhead projector to display a graph from the newspaper or an old textbook. Set a timer for five minutes and invite students to write as much as they can about the graph—for example, "This is a bar graph. It has two columns. It surveyed 88 children. It shows that 42 children prefer creamy peanut butter and 46 prefer crunchy. Four more children prefer crunchy to creamy." When the timer rings, ask children to share what they've written. Record the cumulative responses on a sheet of chart paper. Discuss and categorize the different types of information gathered. Use this chart as a resource the next time you ask children to talk or write about a graph.

TIP

Repeat this technique using a different type of graph. Use students' responses to assess how well they are able to read and understand graphs.

ENGLISH Language LEARNERS

To allow English language learners to focus more fully on the graphing freewrite activity, allow them to dictate their responses as you write them. At the end of the five minutes, read to the student what you have written. Have the child then read it back to you. This process reinforces writing, reading, and oral language skills in a meaningful context.

Weekly Spelling Words Graph

Use this graphing activity to reinforce weekly spelling (or vocabulary) words—and build writing skills.

- Write spelling (or vocabulary) words on index cards. Place them in a pocket chart.

- Invite students to examine the words and think of a way to graph them. For example, words may be graphed according to their part of speech, number of syllables, or spelling pattern. Once the category is determined, use index cards to label the columns of the graph in the pocket chart. Hand out words to students and let them take turns sorting the words into the correct columns on the graph. Repeat the activity using different categories.

Let children share their journals with families at conference time or open-school night to illustrate what they've been learning about graphing.

Graphing Journal

As you study graphing, use journals to move students from talking about graphs to writing about them.

To create the journals, have children cut sheets of graph paper in half and place a dozen or so sheets together, then staple to bind. Have students create a cover for their graph journals, writing their name on it and illustrating with graphing words or pictures of graphs. Throughout the year, students can use their journals to:

- explain the results of a graph
- define graphing terms
- keep track of data
- compare data within a graph
- create different types of graphs using the same information
- record equations that tell about a graph
- compute the mean, median, and mode

Good Morning Lil' Gridders

Use your morning message as a vehicle for discussing and interpreting graphs.

In the morning message, ask students to graph their response to a question that affects the entire class—for example, "What should we study next: amphibians or reptiles?" Include a graph for children to mark responses on the morning message. Add cloze sentences to get students thinking about the graph—for example:

- _____ students want to study amphibians.

- _____ students want to study reptiles.

- We will be studying _____!

When everyone has had a chance to graph their response, let volunteers fill in the blanks using the results of the graph. Encourage students to write additional cloze sentences about the graph for others to complete.

ENGLISH Language LEARNERS

It's always a good idea to give students practice with graphing in a real-world context, such as graphing votes for the next science topic, or the next read-aloud. However, be aware that for English language learners the words may get in the way of the math. To help with this situation, look at graphs for specific vocabulary that may not be familiar to English language learners—for example, in the above activity, the words *amphibian* and *reptile* may be unfamiliar. Make sure you've identified and defined this vocabulary ahead of time. This procedure does double duty. Not only does it ensure that English language learners will get to participate in the graphing activities and share what they know, it will also help them increase their English vocabulary. Support English language learners as they fill in the blanks of the cloze sentences by making number lines available to assist with interpreting the graph.

Story Stacks

Graph students' library books to determine whether they are selecting more fiction or nonfiction.

As students return from the library, ask them to sort their books into two piles, one labeled "Fiction" and the other "Nonfiction." Gather children around the stacks of books. Ask: "Which stack is higher? Does this mean that more fiction or nonfiction was chosen? How can we tell for sure?" Work together with students to count the books in each pile. Create stacks of interlocking cubes to represent each type of book. Ask: "Did you choose more fiction or nonfiction books at the library today? How many more? How many books did the class borrow altogether?" Repeat this activity for several weeks and compare the results. Do students consistently choose one type of book over the other? Ask students to explain their preferences.

FICTION   NONFICTION

Take-Home Activity:
Four-Seasons Pictograph

Connect math, science, and language arts with a graph that invites students to survey friends and family members to find out which season they like best.

Give each student a copy of the poem "Four Seasons." (See page 211.) Read the poem aloud. Discuss the words the poet used to describe each season. Have children take the poem home and share it with friends and family. Instruct them to use the pictograph at the bottom of the page to survey and color in one icon for each person's favorite season. When students return to school with their graphs, ask them to write a paragraph telling why they think one season was favored over another. Encourage them to include their personal thoughts on the season they like best.

Ready, Set, Graph!

Set up a graphing center in your classroom to set the stage for easy and creative graphing experiences.

On a table or low shelf, place graphing materials such as index cards, markers and tape in small baskets or plastic containers. Include a variety of items to use as graphing markers—for example, sticky notes, clothespins, laminated name tags, photo faces (see page 182), interlocking cubes, magnets, or milk caps (from half-gallon or gallon jugs). Provide slips of paper on which students may record their own ideas for graphing questions. Make a pad of chart paper available (or a whiteboard) and invite children to create survey graphs that their classmates can respond to. Use the wall space above the materials to display completed graphs as well as newspaper clippings showing other graphs, charts, tables, or interesting data.

Judy Meagher
Student Teacher Supervisor
Bozeman, Montana

Graph of the Week

When students are involved in the process, adding graphing to your weekly routine is a snap!

At the beginning of the school year, ask each student to think of and record one graphing question on a sentence strip. Questions might include "What color are your eyes? How many people are in your family?" Each week choose one student's question and challenge him or her to design and/or assemble the weekly graph. Let the same student lead the class in a discussion of the graph by asking questions about the data gathered—for example, "How many more students _____ than _____?" This student might also guide the class in calculating the mode and range.

Marianne Chang
Schilling School
Newark, California

Set aside one bulletin board in your classroom or hallway to display your weekly graphs.

Substitute-Teacher Graph

Students are surprised when they find their teacher absent. Help ease the transition by having this graphing activity waiting for your substitute to use.

Label a three-column bar graph with the following categories: Home Sick, At a Meeting, and Out of Town. Label the graph "Where Do You Think Your Teacher Is Today?" Copy a class set of a picture of yourself for students to use as graph markers. (You could also have children draw a picture of you on index cards cut in half.) On the day of your absence, instruct the substitute to have students graph their predictions of your whereabouts. Provide questions for discussing the graph, such as: "How many students think their teacher is home sick? How many more students think their teacher is at a meeting out of town?" Be sure to let the substitute know the reason for your absence so that he or she can share this with students.

Literature LINK

Miss Nelson Is Missing

by Harry Allard (Houghton Mifflin, 1985)

If you're going to be absent, don't forget to leave the substitute a copy of this book. While students ponder your whereabouts, they can hear the story of the kids in Room 207, who long for Miss Nelson's return when they find themselves in the hands of the new substitute, Miss Viola Swamp.

Three-Way Graph

Show students how the same information can look different by varying the intervals on a graph.

Write test scores for an imaginary student on the board—for example, 88, 92, 94, 90, and 92. Ask students to plot the scores on three different graphs using the following intervals:

- by tens from 0 to 100
- by fives from 85 to 120
- by ones from 87 to 95

Invite students to look at the graphs they've created. Ask: "Which graph would you rather show your parents? Which graph looks like your test scores are very poor? Which graph looks like you study a lot sometimes and not at other times?"

Dan Kriesberg
Locust Valley Intermediate School
Locust Valley, New York

Tracking-Time Table

Help students evaluate how they use their time by tracking the number of minutes spent on various activities.

For five days have students use the table on page 212 to track the time they spend on six different activities. They need to make one tally mark for every five minutes. At the end of the week, ask them to record the total number of minutes for each activity in the space provided. Using the data they gather, have them prepare a graph to illustrate how they used their time. Invite students to write a reflective paragraph explaining what they have learned from their tracking-time table and graph. How will this experience affect how they use their time in the future?

Ruth Melendez
High Plains Elementary School
Colorado Springs, Colorado

Collection Graph

Create a graph to help both students and teachers keep track of permission slips or other forms that must be returned to school.

- Collect enough frozen-juice can lids to equal the number of students. Rinse and dry the lids thoroughly.

- Have students glue their photos to the underside of a lid. Attach a piece of magnetic tape to the reverse side.

- Write Yes and No at the bottom of a magnetic board. Line up all the graph markers in the No column. Place a collection box near the graph. As students return their forms, have them move their marker to the Yes column.

- At the end of each day, ask students to check the graph. Use the data to guide a quick discussion that doubles as a reminder. Ask: "How many students have brought in their permission slips? How many more need to bring them in?" Make reminder notes available for students still in the No column.

To create reusable game boards, glue grids to file folders and laminate them. Students can use removable wall adhesive to position treasure chests, and wipe-off markers to record hits and misses.

Treasure Hunt!

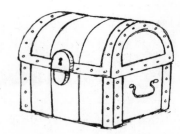

Use this version of the classic game Battleship™ to help students locate points on a coordinate grid.

- To make a game board, tape two coordinate grids (see page 213) to the inside of a file folder (one on each side). Each player will need one game board. Make copies of the treasure chests on page 214, and give one to each player.

- Divide the class into pairs and instruct partners to sit facing one another. Place an object, such as a plastic storage tub, between them (to block each player's view of the other's game board). Have children position their game boards so that the bottom half lies flat on the desk while the top half is propped up against the object between the two players.

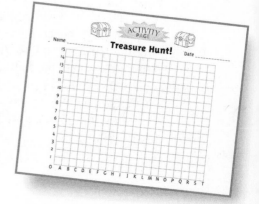

To begin the game, have players position a treasure chest so that it covers 15 squares on the top half of the game board, vertically or horizontally. Players can tape or glue their treasure chests in place or make reusable game boards. (See Tip, page 192.)

Players then take turns calling out points on the grid in an attempt to locate their opponent's treasure. "Hits" and "Misses" are recorded in red and green crayon, respectively, on the bottom half of the game board. The winner is the first player to locate all 16 points that surround their opponent's treasure.

Button Graph

Move students through a series of graphs to help them understand the concepts of concrete, representational, and symbolic. You'll need to have a button for each child to complete the activity.

Use index cards to label columns of a floor graph from 0 to 10. Ask students to stand in the column that shows how many buttons are on the clothes they are wearing. Explain that this is a real graph because we are using real people (ourselves), instead of objects to represent us.

Have students step off the graph and place a photo marker (see Photo Faces, page 182) in the spot where they stood. Define this graph as representational because the pictures look like and stand for us.

Finally, ask students to replace their photo marker with a button. Explain that this graph is called symbolic because the button is a symbol for us, but it in no way looks like us.

On a sheet of posterboard, create a permanent, symbolic graph using the buttons. Discuss the graph and challenge students to calculate the total number of buttons worn by the entire class.

To make a floor graph, purchase a shower curtain from a discount store. Create a grid on the curtain using strips of masking tape. As students participate in other graphing experiences, ask them to identify the type of graph as concrete, representational, or symbolic.

Class-Photo Graph

Show students how to use a Venn diagram to organize information about themselves and others.

blue eyes   wears glasses

⑤ To prepare for the activity, record students' physical characteristics on index cards—for example, blue eyes, curly hair, or wears glasses. Glue student photos to small squares of cardboard or clean frozen-juice can lids. Draw a Venn diagram on a sheet of posterboard and laminate.

⑤ Choose two index cards, such as brown eyes and curly hair. Tape them beneath the two circles. Work together with students to sort the photo markers into the different categories. A student who has both brown eyes and curly hair will place his or her photo in the space where the two circles intersect. A student who has neither brown eyes nor curly hair will place his or her photo in the space outside the two circles.

⑤ Display the Venn diagram and ask, "How many students have brown eyes? How many students have curly hair? How many students have both brown eyes and curly hair? How many students have neither brown eyes nor curly hair? In which category are the most students? The fewest?" Repeat the activity several times using different physical characteristics.

Check for food allergies before letting children eat their cereal graph markers.

Cereal Snack Graph

With this colorful graph, students learn how to write equations to represent data.

Give each child a small bag of multi-colored O-shaped cereal pieces and a sheet of large-square graph paper. Show students how to sort their cereal by color, as shown. Based on the total number of cereal pieces in each column, have them write equations for:

⑤ blue + red (repeat using other color combinations)

⑤ the three primary colors (red + blue + yellow)

⑤ the two colors that make green (blue + yellow)

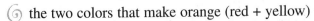

- the two colors that make orange (red + yellow)
- the two colors that make purple (red + blue)
- the colors of grass and sky (green + blue)
- the three colors of a traffic light (red + yellow + green)
- their two favorite colors

Invite students to transform their real graphs into symbolic ones. Have them remove (and eat) their cereal pieces one at a time and color in the squares to replace each one accordingly.

Change the color equations to reflect the cereal you use.

Kelley Foster
Cicero Elementary School
Cicero, New York

Take-Home Activity:
Graphing Goes Home

Help parents reinforce what their children are learning in school with a take-home letter that lists fun ways to explore graphing at home.

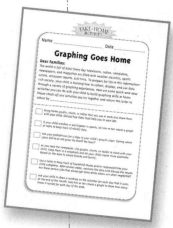

Give each child a copy of the take-home letter on page 215. Read through the checklist together to review ways they can learn about graphing at home. Ask children to check off the activities as they try them. They can set a personal goal for the number of activities, or you may recommend a number. As students return the checklists to school, encourage them to bring in samples of the work they completed to share with classmates. Display their at-home graphing projects on a bulletin board with a sign that reads Graphing Goes Home!

TIP

You may want to add food coloring to the water to make the graph easier to read.

Pick Your Pop!

Create a graph using two-liter bottles to determine which type of soda students like best.

- Collect six empty 2-liter bottles. Label each with a different flavor of soda: root beer, cola, orange, grape, lemon-lime, and cherry.

- Set the bottles on a table and ask, "Which type of soda would you pick?"

- Let students cast their vote by pouring a half-cup of water into the bottle that represents their favorite. Use a funnel to avoid spills.

- Draw students' attention to the height of water in each bottle. Let students tell you which type of soda is the most popular and which is the least popular.

- Ask: "How can we determine the number of students that voted for each type of soda?" To do this, pour the contents of each bottle into a large, glass measuring cup. Double the number of cups to determine the number of students.

Take-Home Activity:
Graphing by the Slice

Based on a survey of favorite pizza toppings, students create circle graphs to display the results.

- Send each child home with a copy of the pizza slices and toppings. (See page 216.) Instruct students to survey six friends and/or family members to find out which pizza topping they like best: plain cheese, pepperoni, mushroom, sausage, or veggie. For each response, have them cut out the topping and glue it to a slice. For plain cheese responses, slices should be left as is.

- Once students have all six responses, have them cut out the slices, sort them by topping, and glue them in a circle formation to another sheet of paper. They can use extra copies of the toppings to create a key for the graph.

- Back at school, help children record fractional amounts and/or percentages to show how many people preferred each topping. Display the circle graphs on a bulletin board with a sign that reads "Graphing by the Slice."

Coordinate Twister

In this adaptation of a favorite floor game, students use their hands and feet to plot coordinates on a giant grid.

⊙ Use strips of masking tape to create a grid on the classroom floor that has 20 squares and is approximately 4 by 5 feet. Use index cards to label the horizontal axis with letters (A, B, C, D, E) and the vertical axis with numbers (1, 2, 3, 4).

⊙ Use index cards to make a set of body part cards that include right foot, left foot, right hand, and left hand. Use a different color of index cards to make coordinate cards in all of the possible combinations: A, 1; A, 2; A, 3; A, 4; B, 1; B, 2; B, 3; etc.

⊙ Shuffle both sets of cards. Choose two players and one caller. Have the caller stand with his or her back to the grid and select a body part and coordinate card to give each player a movement command. For example, a player might have to place her left foot on E, 4. When a player is unable to follow a command or remain upright, he or she is replaced by another player and a new round begins.

Literature LINK

The Fly on the Ceiling

by Dr. Julie Glass (Random House, 1998)

Introduce students to coordinate grids with this humorous math myth. This book tells the story of French philosopher René Descartes and how he discovered the Cartesian Coordinate system while looking for a way to organize all his "stuff"!

TIP

When test time rolls around, let students flip through their graphing journals (see page 186) to review and explain what they've learned about graphing.

Data Detectives

Challenge students to search through books, newspapers, magazines, and the Internet for interesting information suitable for graphing.

- Show students examples of data you've discovered in the media—for example, daily temperatures from cities across the country or sports scores. Work together to create graphs from this data to serve as models for students' own work.

- Have students search for data at home, using any of the resources mentioned. (You might also make copies of the daily paper available for students to borrow.) Ask students to clip or copy the information (being sure to record the data source), and bring it to school.

- Help students decide which type of graph (line, bar, circle, etc.) would best represent the information they've found. Supply markers and posterboard and invite them to create graphs based on their data. Remind students of the models they created earlier and encourage them to use them as resources.

- Provide time for students to share their graphs with classmates. As they do, ask them to identify the data source and to explain why they chose a particular type of graph.

Literature LINK

The Factastic Book of 1001 Lists

by Russell Ash (Dorling Kindersley, 1999)

This fact- and figure-packed book is a great source of graphing data. Look for information on everything from animals to the arts. Kids can graph the ocean depth, high (and low) temperatures, wingspan, sound levels, building height, sports scores, and the measurement of each planet.

TIP

Students can use programs such as *Data Wonder* (Addison Wesley), *TableTop Jr.* (TERC), and *The Graph Club* (Tom Snyder) to create graphs on the computer.

TIP

USA Today is a great source of colorful graphs, charts, and tables.

Graphing in the News

In this activity, students make real-world connections as they read and interpret graphs in the newspaper.

Invite students to scan the newspaper for several days until they find a graph that interests them. Have them glue the graph to a sheet of paper and write the answers to the following questions:

- What is the title of this graph?
- What does this graph tell you? Name three things.
- Who might benefit from this information?
- What type of graph is used to represent the data?
- What is the source of this information?

Display the graphs on a bulletin board and let students compare and contrast them.

TIP

As an alternative activity, have students write five questions about their graph beside it on their paper. Let students exchange papers and answer one another's questions.

Take-Home Activity:
Coordinate Maps

For an activity that is sure to hit home, invite kids to use coordinate grids to create a map of their bedroom. To first model this assignment, work together with students to create a coordinate map of the classroom or playground.

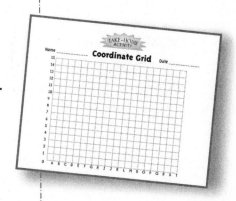

- Send each student home with a copy of the coordinate grid. (See page 217.)

- Challenge students to create a map of their bedroom by plotting only those objects that take up floor space, such as a bed, dresser, bookshelf, chair, or toy box. Have them write the name and/or draw a picture of the object next to each point.

- On a separate sheet of paper, ask students to write five questions based on their map—for example, "Where is my bed? What object can be found at point (6, 5)?" Invite students to exchange papers and answer one another's questions.

Monday's Child

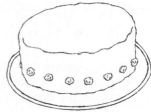

Inspired by a traditional rhyme, students investigate and graph the day of the week on which they were born.

○ Share the poem "Monday's Child" with students. (See page 218.) Have them find out the day of the week they were born on and fill in the last line of the poem.

○ Make seven extra copies of the cake pattern and label them with the days of the week. Glue the cakes to a large sheet of craft paper. Give each child a birthday candle. Have children glue their candles to the cake that shows their birthday. Gather children around the graph. Count the number of students born on each day. Ask questions such as:

- On which day of the week were the most (and fewest) students born?

- How many more students were born on Monday than Tuesday?

○ Reread the poem, then ask: "Which line tells about you?" Invite students to tell whether they think their line accurately depicts them. Make new poems by letting children rewrite the line of the poem so it accurately reflects their personality and interests—for example, "Monday's child is good at sports."

Bottles and Cans Pictograph

Let students create a pictograph to show the number of bottles (plastic) and cans they collect during a school-wide charity drive.

○ Plan a recycling drive to raise money for a charity of your students' choice. Have students work together to create flyers and posters to advertise the event. Place a large garbage can in an accessible location to house the donations.

TIP

Students may be able to find out on which day of the week they were born by asking their parents. They can also find this information on the Internet at **highhopes.com/ 21centurycalendar. html**.

Collect for five days. Near the end of each day, bring the bottles and cans back to the classroom. Work with students to tally the total number of each. Provide students with gloves to handle the donations. (The school nurse may be able to provide disposable gloves for this.)

When the week is over, help students create a pictograph to show how many bottles and cans were collected each day. Make multiple copies of the bottle and can templates on page 219. Have students color them and cut them out. To make the graphs (one for bottles, one for cans), label the horizontal axis with the days of the week that correspond to the drive. For each day, have students cut and paste one bottle or can for every two collected. If the total number is odd, show students how to cut a bottle or can in half to represent one bottle.

Post the completed graphs and ask questions to guide a discussion: "On Monday, did we collect more cans or bottles?" (repeat for other days of the week) "On which day did we collect the most cans? Bottles? Both? How many cans did we collect altogether? Bottles? Both? How much money is each bottle/can worth? How much money will the bottles raise for charity? Cans? Both?"

Extend the activity by creating a line graph to show how much money is collected from the drive each day. Display the graphs in the school lobby with a big sign saying "Thank You!" Take the cans/bottles to a recycling center and donate the proceeds to charity.

And the Survey Says . . .

Invite students to create and conduct surveys to learn more about fellow classmates.

Copy your class list on a survey form and make a copy for each student. Ask each student to come up with a survey question and two to five suitable responses—for example, "How much sleep do you get on a weeknight: 8 hours or less, more than 8 hours but less than 9 hours, from 9 to 10 hours, more than 10 hours?"

Have students work over a period of a few days to gather responses from each of their classmates, then summarize the data in a graph. Put their graphs together to make an informative class book!

Trina Licavoli Gunzel
Lincoln Elementary School
Corvallis, Oregon

You may want to begin with a concrete graph using the real bottles and cans. Then, for every two bottles/cans you remove from the concrete graph, glue one bottle/can to the pictograph.

Take-Home Activity:
Graphing Goes to Work!

How are graphs used in the workplace? Challenge students to find out!

- Ask students to survey parents, grandparents, and friends of the family to discover how they use graphs, charts, or tables in their work. Encourage them to bring in at least one example to share with the class.

- As students share their examples, encourage them to identify the line of work and explain how the graph, chart, or table is used.

- Use these workplace models to create a bulletin board display. Randomly scatter sheets of graph paper over colorful bulletin board paper to create a backdrop for the examples. Label the display "Graphing Goes to Work!"

Before-and-After Graph

Planning an insect unit? As an introductory activity, use a scatter graph to assess children's attitudes about insects.

- To create the graph, make two copies of the ladybug pattern on page 220. Write Yes at the bottom of one and No on the other. Tape the ladybugs to the board under the question, "Do You Like Insects?"

- Give each student a black sticky dot. Ask students to graph their response to the question by placing their dot on the corresponding ladybug. Together, count up each ladybug's spots to determine how many students do or do not like insects.

- Readminister the graph after your insect unit and compare the results. Have attitudes changed? Ask students to reflect on why or why not their thoughts about insects might be different.

Wendy Weiner
The Parkview School
Milwaukee, Wisconsin

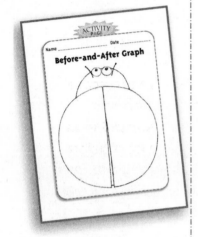

Try other before-and-after graphs with students as you study bats, spiders, or snakes!

Shape Scatter Graphs

Let students learn more about scatter graphs with an activity that lets them survey parents at open-school night or parent conferences.

- Ask each student to think of a graphing question with no more than three responses—for example, "What kind of ice cream do you like: chocolate, strawberry, or vanilla?"

- Show them how to create a shape graph by drawing and cutting out a large shape or shapes that correspond to their question—for example, an ice cream cone with three different scoops for the above question.

- Ask students to write the question on a sentence strip and hang it along with their scatter graph at a display. During open-school night or at conferences, give parents a sheet of sticky dots. Invite them to use the dots to respond to each graph.

- The next day, ask students to write a few sentences telling about the data they've collected on their shape graph.

Judy Meagher
Student Teacher Supervisor
Bozeman, Montana

Guide students in making several scatter graphs to respond to as a class before assigning them to create their own.

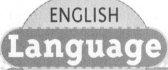

You may want to have English language learners dictate their survey questions for you to record. After writing them on the form, let the students read back the questions. This will strengthen word recognition skills in meaningful ways and help children participate more fully in the activity. If any of your English language learners need assistance reading classmates' names, give them a blank survey form and have them ask students to record their name when asked the survey question. They can read the name back to themselves, reinforcing spelling and pronunciation.

Garden Graph

Use a line graph to compare the growth of two plants, and spark spin-off explorations on seeds and plants.

- Purchase two different bulbs (for example, amaryllis and narcissus) from a garden store. Let students plant them in pots and place them in a suitable spot (indoors) for growing.

- Label the horizontal axis of a line graph with the number of weeks (10) and the vertical axis with the number of inches (20). Each week measure and record the height of each plant using the line graph.

- Use two different-color markers to distinguish between the two plants. Once both bulbs are fully grown, discuss the graph:

  - How many weeks did it take for the amaryllis bulb to bloom? The narcissus bulb?

  - How many inches did the amaryllis grow altogether? The narcissus?

  - Which bulb grew the fastest? The tallest?

  - Looking at the data for the amaryllis [narcissus] bulb, which week showed the most growth? The least growth?

Charlotte Sassman
Alice Carlson Applied Learning Center
Forth Worth, Texas

TIP

Take the activity further by letting children work in teams to design a similar experiment to track and graph growth of other plants.

Graph a Snack

Show students how to use a graph to compare the fat content of their favorite snacks. Use what students learn to guide lessons on nutrition.

- Collect mini-size (1 oz) empty commercial snack bags that students bring in their lunch, such as those for potato chips, pretzels, or cheese twists. Once you have duplicate bags for four or five different kinds of snacks, create a line graph by gluing one set of the bags (one of each type) to the horizontal axis of a graph. Label the vertical axis from 0 to 15 to represent the number of fat grams per bag.

- Pass out the extra bags you've collected and show students how to read the nutritional information on the package. Explain that *g* is an abbreviation for grams and help them locate the number of fat grams in each bag. Provide nutritional information to help students discover how much fat they need in their diets each day.

- ⟲ Record the fat grams for each type of snack on the graph. Ask: "How many grams of fat do the pretzels have?" (Repeat for each type of snack.) "Which snack contains the greatest amount of fat? Least? How does the amount of fat in your favorite snack compare with the amount of fat you need each day? Which snack is the healthiest? Why?"

Elizabeth Wray
Blair Middle School
Norfolk, Virginia

Grow-a-Beast Graph

Use attention-getting "grow beasts" for an investigation that lets students create line graphs to record and compare the data they collect.

- ⟲ Purchase two identical grow beasts at a craft or discount store. Show them to students and ask them to predict whether the beasts will grow larger in salt water or distilled water.

- ⟲ Place the beasts in separate containers of the same size. Fill each container with a different type of water (distilled or salt) and place them in the same area of the room to control for light, airflow, and surrounding temperature.

- ⟲ At the same time each day, remove the beasts from the containers. Drop them from a height of one foot to knock off excess moisture. Use a balance scale to obtain the mass in grams of each beast. Set up a table to record the data.

- ⟲ Repeat the process for 12 days. At the end of this period, ask each child to create a line graph showing time versus mass for each beast. Use the graphs to discuss the results of the experiment. Ask children why they think the distilled-water beast grew larger than the salt-water beast. (The answer is related to the concept of *diffusion*—when particles from two or more substances intermingle. In the plain water, there are fewer particles inside the "beast" than outside so the water is free to move into the creature, making it larger. In the salt water, there are more particles outside the beast. The water will move into the beast until there is an equal amount outside and inside.)

Pamela Galus
Burke School
Omaha, Nebraska

Encourage students to create similar graphs based on other nutritional data such as calorie or carbohydrate counts. For nutritional information, including the recommended daily allowances for fat and other nutrients, go to **www.usda.gov**.

Dinosaur Data

Given a set of clues, students uncover and graph the heights of five dinosaurs.

🌀 Give each student a copy of the Dinosaur Data sheet. (See page 221.) Ask children to read each clue at the top of the page. Show them how to use the information to calculate and graph each dinosaur's height.

🌀 Encourage students to look at the completed graph and make other comparisons among the dinosaurs—for example, how many spikes, length of tails, and so on. Students can make visual representations of the comparisons for a fun display.

Literature LINK

How Big Were the Dinosaurs?

by Bernard Most (Harcourt Brace, 1994)

This book invites children to make comparisons between the size of dinosaurs and everyday objects, such as a school bus. For more comparisons, see *The Littlest Dinosaurs* (Harcourt Brace, 1989), also by Bernard Most.

Graphing Animal Data

Science is full of interesting facts and figures! During your next animal unit invite students to create graphs using amazing animal data.

🌀 Divide the class into six groups. Give each group a different animal data card. (See page 222.) Challenge children in each group to work

TIP

For more dinosaur fun take your students to **kidsdomain.com /games/dino. html**, where they can play a coordinate game called "Dinosaur Dig."

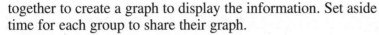

together to create a graph to display the information. Set aside time for each group to share their graph.

⑤ Encourage them to ask each other questions—for example, "Why did you choose to display your data on a bar graph? What was the greatest challenge for you in creating this graph?" Display the graphs in the hallway for other classes to enjoy.

Weather Watchers

Help your young meteorologists track and graph the weather using the local newspaper.

⑤ Make copies of the newspaper weather page for each student. Draw students' attention to the various types of data shown, such as temperature highs and lows, rain and snowfall amounts, wind speeds, and times of sunrise and sunset.

⑤ Each month choose one type of data to collect. As part of your morning routine, record the previous day's data on your calendar grid, using the newspaper as a resource. At the end of the month, help students decide on the best type of graph to display the information collected. Let pairs of children lead the class in creating the graph each month. Place graphs in a binder to create an almanac so that students can easily revisit the graphs and build on their understandings.

 ## Musical Pictograph

After listening to musical selections, students graph the type of music they enjoy most.

⑤ Choose five musical selections, one each from the following categories: jazz, classical, country, rock and roll, and new age. Ask students to listen attentively as you play each song.

⑤ Replay a small portion of each song and invite students to stand up when they hear the type of music that best represents their preference. Create a table on the board to record the number of students for each musical category. Work together with students to create a pictograph of their musical preferences.

Not Your Average Tail!

In this activity, students calculate the mean, median, mode, and range of six lengthy animal tails.

How Long Are Their Tails?

Asian Elephant:
59 inches

Leopard:
55 inches

African Elephant:
51 inches

African Buffalo:
43 inches

Giraffe:
43 inches

Red Kangaroo:
43 inches

🌀 Divide the class into six groups. Assign each group a different animal. (See How Long Are Their Tails?, left.) Using various art materials and the measurements given, instruct each group to create the animal's tail and record its length on it.

🌀 Display the tails vertically on a wall or bulletin board. Make sure they are even at the top. Challenge students to determine the range, mean, median, and mode of the tails.

Take-Home Activity: Hidden-Picture Graph Packs

With this take-home activity, students plot and connect points to reveal a hidden picture.

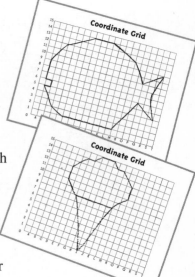

🌀 Make one copy of the coordinate cards (see page 223) and cut them apart. Place each one in a resealable, plastic bag along with crayons and a copy of the coordinate grid. (See page 217.)

🌀 Show students how to use the cards to plot and connect the points in order on the grid. Encourage them to personalize their pictures by adding color and details.

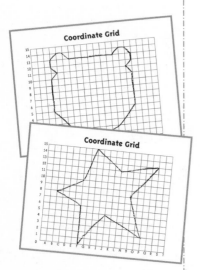

🌀 Make the picture packs available for students to sign out on a daily or weekly basis. Invite students to create their own picture packs to share with classmates. To do so, they should first draw a design on the grid and then list the coordinate pairs on an index card.

Deborah Rovin-Murphy
Richboro Elementary School
Richboro, Pennsylvania

Graphs on the Move

With this physical graph, students use movement to respond to the question "During which month were you born?"

- Read aloud "Birthdays." (See page 224.) As you do, invite students to reveal their birth month by responding with the movements named in the chant. For example, children born in September will touch their toes.

- Read aloud the chant a second time, this time having a volunteer "data collector" tally the number of students for each month. (Be sure this child counts himself or herself, too.) Create a bar graph to display the data collected.

- Fill in the squares in each column using crayons in colors that match each month's birthstone. (See Birthstone Colors, below.)

Birthstone Colors

Month	Stone	Color
January	Garnet	Red
February	Amethyst	Purple
March	Aquamarine	Light Blue
April	Diamond	Clear or Yellow
May	Emerald	Green
June	Pearl	White
July	Ruby	Red
August	Peridot	Light Green
September	Sapphire	Blue
October	Opal	White
November	Topaz	Yellow
December	Turquoise	Green-Blue

Name _____

Date _____

Look What I'm Reading!

Historical Fiction

Realistic Fiction

Folklore

Biography

Mystery

Sci Fi/ Fantasy

Informational

Poetry

The Great Big Idea Book: Math © 2009, Scholastic Teaching Resources

Name _____

Date _____

Four Seasons Pictograph

Four Seasons

Spring is showery, flowery, bowery.
Summer: hoppy, choppy, poppy.
Autumn: wheezy, sneezy, freezy.
Winter: slippy, drippy, nippy.

— Anonymous

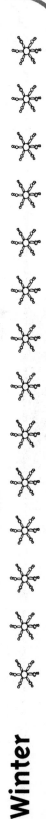

Spring

Summer

Autumn

Winter

Name _____

Date _____

Tracking-Time Table

Activity	Amount of Time					Weekly Total
	Monday	Tuesday	Wednesday	Thursday	Friday	
Doing Homework						
Watching TV						
Playing Sports						
Playing						
Doing Chores						
Using a Computer						

Treasure Hunt!

Name _____

Date _____

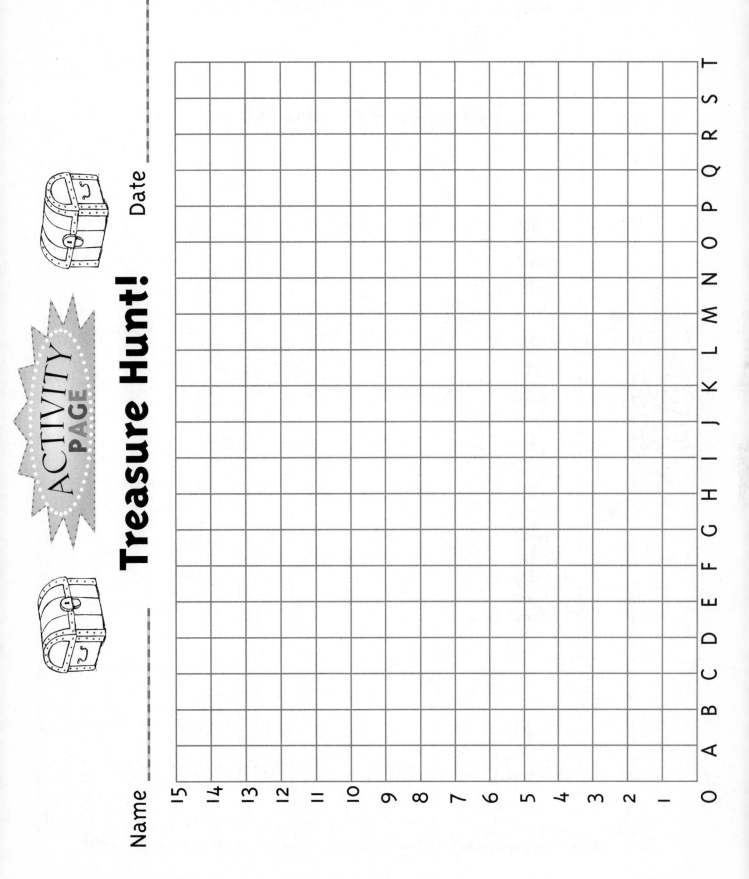

Name _____   Date _____

Treasure Hunt!

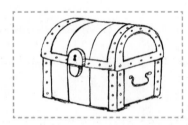 

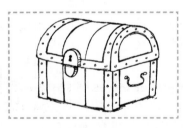 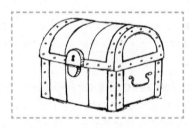

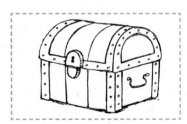 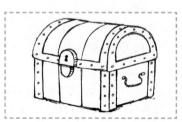

Name _____ Date _____

Graphing Goes Home

Dear Family,

The world is full of data! Every day televisions, radios, computers, newspapers, and magazines are filled with weather statistics, sports scores, consumer reports, and trivia. To prepare for life in this information-rich society, your child is learning how to collect, display, and use data through a variety of graphing experiences. Here are some quick and easy activities you can do with your child to build graphing skills at home. Please check off any activities you try together and return this letter to school by _____.

☐ Bring home graphs, charts, or tables that you use at work and share them with your child. Discuss how these tools help you in your job.

☐ If your child watches or participates in sports, let him or her create a graph or table to keep track of related data.

☐ Ask your pediatrician for a copy of your child's growth chart. During which years did he or she grow the most? The least?

☐ As you read the newspaper, clip graphs, charts, or tables to share with your child. Keep them in a scrapbook and let your child create trivia questions based on the data, to amuse friends and family.

☐ Use a table to keep track of household chores and/or responsibilities your child completes. After several weeks, examine the data and discuss the results. Are there certain jobs that always get done while others are often neglected?

☐ Ask your child to draw a raindrop on the calendar for each day that it rains. At the end of the month, help him or her create a graph to show how many days it rained each week.

Name _____ Date _____

Graphing by the Slice

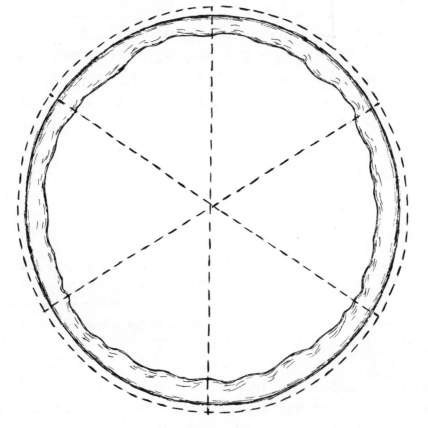

pepperoni

mushroom

sausage

veggie

Name _____

Coordinate Grid

Date _____

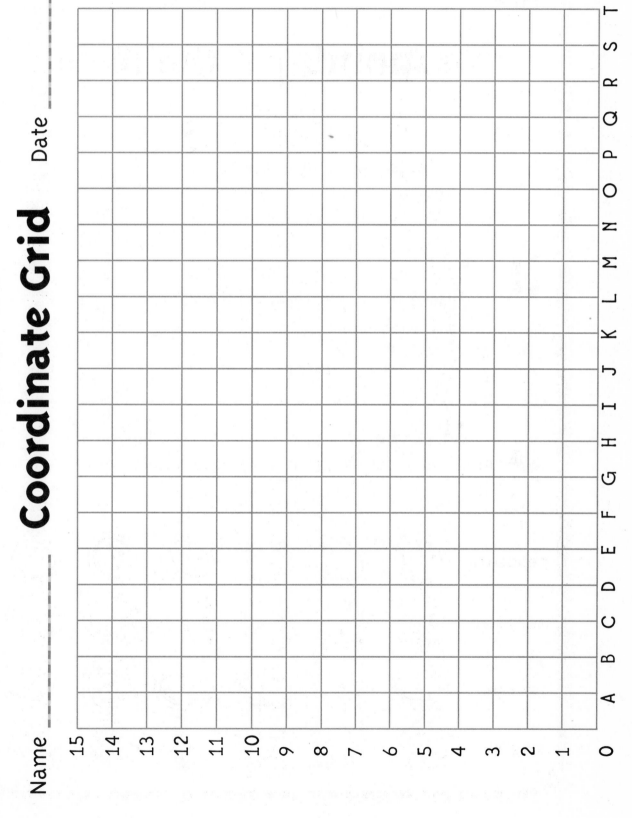

Name _____ Date _____

Monday's Child

Monday's child is fair of face,

Tuesday's child is full of grace,

Wednesday's child is nice to know,

Thursday's child is on the go,

Friday's child is loving and giving,

Saturday's child makes life worth living,

And the child who is born on the seventh day,

Makes many friends along the way!

I was born on a _____!

—*adapted by Jacqueline Clarke*

Color candles on the cake to show how old you are.
Write in the name for the day of the week on which you were born.

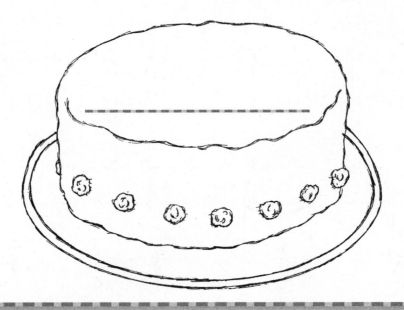

Name _____   Date _____

Bottles and Cans Pictograph

Name _____ Date _____

Before-and-After Graph

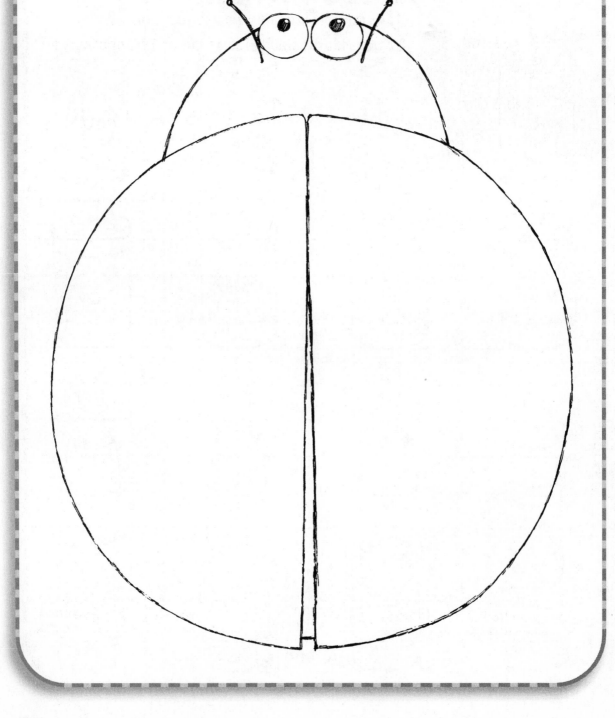

The Great Big Idea Book: Math © 2009, Scholastic Teaching Resources

Name _____   Date _____

Dinosaur Data

Clues

1. Tyrannosaurus Rex was 50 feet tall.

2. Ultrasaurus was twice as tall as Tyrannosaurus Rex.

3. Diplodocus was 10 feet shorter than Ultrasaurus.

4. It would take three Stegosauruses to equal the height of Diplodocus.

5. Stegosaurus and Triceratops were the same height.

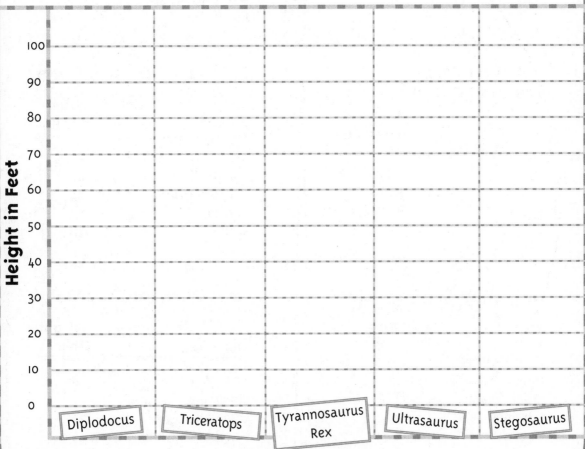

Height in Feet (y-axis: 0, 10, 20, 30, 40, 50, 60, 70, 80, 90, 100)

Diplodocus | Triceratops | Tyrannosaurus Rex | Ultrasaurus | Stegosaurus

Type of Dinosaur

Name _____   Date _____

Graphing Animal Data

How Long Can They Stay Underwater?

Mammal	Number of Minutes
Human	1
Sea Otter	5
Platypus	10
Hippopotamus	15
Seal	22

How Long Do They Sleep?

Animal	Average Hours of Sleep per Day
Koala	22
Two-Toed Sloth	20
Little Brown Bat	19
Giant Armadillo	18
Child	10

How Fast Does Their Heart Beat?

Animal	Heartbeats Per Minute
Bat	750
Cat	120
Sheep	75
Horse	40
Frog	30

How Fast Can They Swim?

Animal	Miles Per Hour
Sailfish	68
Bluefin Tuna	46
Yellowfin Tuna	44
Blue Shark	43
Wahoo	41

How Long Can They Live?

Animal	Years
Giant Tortoise	150
Human	121
Killer Whale	90
Common Toad	40
Queen Ant	18

How Much Do They Weigh?

Mammal	Weight in Tons
Blue Whale	128
Fin Whale	44
Gray Whale	32
Humpback Whale	26
Pilot Whale	3

The Great Big Idea Book: Math © 2009, Scholastic Teaching Resources

Hidden-Picture Graph Packs

Top-left pack

1. (H,15)
2. (G,14)
3. (F,14)
4. (D,12)
5. (D,10)
6. (F,8)
7. (I,0)
8. (L,8)
9. (N,10)
10. (N,12)
11. (L,14)
12. (K,14)
13. (J,15)

Top-right pack

1. (F,13)
2. (E,14)
3. (D,14)
4. (C,13)
5. (C,12)
6. (D,11)
7. (C,10)
8. (C,5)
9. (F,2)
10. (H,1)
11. (K,1)
12. (M,2)
13. (P,5)
14. (P,10)
15. (O,11)
16. (P,12)
17. (P,13)
18. (O,14)
19. (N,14)
20. (M,13)

Bottom-left pack

1. (G,14)
2. (C,12)
3. (A,9)
4. (A,6)
5. (B,6)
6. (B,5)
7. (A,5)
8. (C,2)
9. (E,1)
10. (M,1)
11. (P,6)
12. (S,4)
13. (R,7)
14. (R,8)
15. (S,11)
16. (P,9)
17. (N,12)
18. (J,14)

Bottom-right pack

1. (H,15)
2. (F,10)
3. (A,8)
4. (F,6)
5. (F,0)
6. (J,5)
7. (P,2)
8. (N,8)
9. (R,13)
10. (L,12)

Name _____   Date _____

Birthdays

Hey, hey
When's your birthday?

Clap your hands
If it's January

Stamp your feet
If it's February

Shrug your shoulders
If it's March

If it's April
Up you stand

Born in May
Wave your hand

June's the month
To touch the sky

Fly around
If it's July

If it's August
Blow your nose

In September
Touch those toes

If your day is in October
Start that day
By rolling over

In November
Bend your knees

Here's December
You must freeze!

— *Sonja Dunn*